THIS LUMINOUS COAST

THIS LUMINOUS COAST

JULES PRETTY

FULL CIRCLE EDITIONS

Also by Jules Pretty

The Earth Only Endures — *2007*
Agri-Culture — *2002*
The Living Land — *1998*
Regenerating Agriculture — *1995*

Fertile Ground — *co-author, 1999*
Trainers' Guide to Participatory Learning and Action — *co-author, 1995*
The Hidden Harvest — *co-author, 1992*
Unwelcome Harvest — *co-author, 1991*

Sustainable Agriculture and Food — *ed. 2008*
Sage Major Work on the Environment — *ed. 2006*
The Pesticide Detox — *ed. 2005*
Earthscan Reader in Sustainable Agriculture — *ed. 2005*
Guide to a Green Planet — *ed. 2002*

Nature and Culture — *co-editor, 2010*
Sage Handbook on Environment and Society — *co-editor, 2007*

For Gill, Freya and Theo
& Chris

The land is always stalking people
The land makes people live right

Annie Peaches [Western Apache] Arizona,
in Keith Basso, *Wisdom Sits in Places*

Contents

Photographs by the author

ESSEX
SUFF
W
S
N
E

NORFOLK
K

Preface
A Year on the Coast

This is a coast about to be lost. Not yet, but it will happen soon. A thousand years ago, a king showed his courtiers how absurd it was to command the waves to retreat from this shore. Others built sea walls and estuary defences around the whole of the region. Small stretches of cliffs provided natural protection, as did shingle heaped up by the sea. Revetments have been added, and sea walls raised, and even though they mostly did their job, with some long-remembered exceptions, churches, houses and some whole settlements fell into the sea. The common-field system came and went, then the agricultural revolution, and lately came the industrial age. We thought we were in control, with the means to remain an island of a size largely unaltered since the last ice age. There were siren calls, but we ignored them. Until this last generation, when the dots were finally joined.

Fossil fuels that drive our industrial economy, which in turn brings so much, have polluted the atmosphere with carbon dioxide and other waste gases. These absorb reflected light from the earth and warm up, and more atmospheric energy provokes climate change. A warmer world also makes water expand. And for the 70 per cent of the earth's surface covered by oceans and seas this means one thing: more than a billion cubic kilometres of water have to go somewhere. And that is upwards onto the land and its beaches, marshes, dunes, mudflats, grazing meadows and shingle banks. All are now under threat. And as if this is not enough, East Anglia is sinking too. It seems doubly unfair, but since the glaciers retreated from northern Britain, the land there has been bouncing upwards and thus levering down the south-east.

Some predictions are gloomy. In 50 to 100 years, perhaps no landscapes by the sea will survive quite as they are today. What, some may ask, is there to lose? The interface between land and sea is ever-changing: it will adapt, as will the wildlife. We can, as it were, manage the retreat. We may lose a remote house or two on a cliff, a slick and grimy marsh, a Victorian waste tip, a windswept grassland peppered with skinny sheep. Maybe a beach will lose its sands, but how often does the sun shine on a North Sea shore anyway? Besides, our clever industrial age will come up with a fix soon enough, and all those doom-mongers will have to eat their words.

I took out a large-scale map to look more closely at the three counties of Essex, Suffolk and Norfolk, otherwise known as East Anglia, and traced a line from south to north, measured distances, and realised two things. Most of the shoreline was in easy reach of where I lived, where I had grown up too, yet surprisingly there were many places I had never been. I knew them by reputation or from books, but not first hand. It was clear, too, from the map that much of the character of the region must be defined by its proximity to the sea. It's surrounded on three sides, intercut by rivers, suffused by the light off the water. Well-known wild places and natural landscapes sit side by side with settlements, power stations, military installations and clanging ports. I wondered, how would this rim of land by the sea look if you walked from one end to the other? Sometimes an idle thought is like a retrovirus: it gets stuck in your DNA and replicates. I came to realise I should walk the coast before it was lost. Perhaps this would be no more than a lament, but maybe too there would be the undiscovered in the near by.

I bought more detailed maps, cut them up, stuck pieces together, and started taking notes about the named places of the coast. And also notes on the gaps, the apparently empty quarters. I came to realise there were many but of course none was empty. I built a small library of old books with faded dust jackets, and new ones that shined. I searched bookshops and websites for the discontinued ones, and they came musty, matured, smelling of ancient farmhouses. My first idea was to make one long walk, a single grand expedition, but this plan would later change. These three seaward counties of the East Angles and East Saxons were once joined to Continental Europe, when the last ice age locked up so much ocean water in ice. Now the coast strongly defines this region. Beach, salting, seawall and marshland. Fishing and smuggling, farming and sailing. Birds watched and birds shot. Created communities, deserted resorts, eroded cliffs, villages underwater, caravan parks and whole new invented places. All on a linear stretch of land and sea hundreds of miles from the Thames Estuary at the east of the capital to the Wash at King's Lynn. The more I read, the more I realised that I'd already started, anticipating what might come, and so walking a future memory.

How far would it be? It's one of the first questions for any expedition. Yet coastlines are famously fractal, just like tree surfaces, as Richard Mabey has observed. How long really is a coast? Do you measure around every small cliff or promontory, around every stone on the beach, or even every grain of sand? Walking takes you around more twists and turns than a line drawn on a large-scale map. Do you measure the high tideline, or the shorter low one? Which islands and rocks should be included? And where does the salty coast end and freshwater rivers begin? On some estuaries, imposing flood barriers separate salt from fresh water. But on others, the shore and river are constantly intermixed, changing with every tide. In truth, the coast and shore is not really a line. It's more of a zone. At a guess, there are 400 miles or so in Essex, with its five long estuaries; then 220 in Suffolk, of which only 50 are by the sea, and the rest on tidal shorelines in another five rivers; and 70 or so in Norfolk, mainly because few rivers penetrate its barrier coast. Maybe 700 miles in all? But far fewer if you only count the parts that face the open sea. The simple plan was becoming more complicated.

Another uncertainty lies in the region's name and thus identity. East Anglia itself is a slightly problematic term. On large maps, it tends to include the lowland Fens and shire counties of Hertford, Bedford, Northampton and Cambridge, as well as the three coastal counties of Norfolk, Suffolk and Essex. But this is strictly larger than the kingdom of the East Angles, which did not include Essex. And Essex now as a county does not include swathes of eastern London and recently established unitary authorities, even though people there still feel they reside in Essex. East Anglia seems to cover Essex, Suffolk and Norfolk reasonably well, as that is how we generally understand this bump of land on eastern England today. It is curious that these counties should even have different identities. The borders are, after all, only administrative lines on the map, ancient though they are. One of my aims, therefore, was to walk the whole coast and its communities and ecologies, and learn what I could about the specificities of place.

I knew that this would also feel partly like my land, as my mother's side of the family lived in the east of London and then migrated out to south Essex, and my father's side had long been resident in Suffolk and north Essex. I had been native to some places, a bedroom looking east over the North Sea; others I had never been to at all. Some I knew from only certain times of my life, but never visited since. A further idea, therefore, was to create some kind of memory map of this coastline. The journeys would explore words and images in a remapping of the coast, in which I intended, as Marina Warner recommends, to listen to other people's ghosts as well as my own. An aim was to interweave stories of the land and sea with people past and present. I would come to meet and listen to stories from oystermen, wildfowlers, reed-cutters, lifeboatmen, barmaids, farmers, solicitors, nature wardens, writers, artists, ferrymen, priests, businessmen, and a whole lot of taxi-drivers.

Once, our ancestors walked the world. Then came domestication of animals, the wheel, and now the car. Today walking can be hard, as settlements and transport have become rearranged beyond our control. Many people still walk for pleasure, but few of us now walk far as part of our daily lives. Ronald Blythe remembers that the region's footpaths were once full of people moving about, working, interacting. These were like today's main roads, except people talked and walked and watched. The old countryside was peopled. Blythe writes, "Friends never tire of telling me that my life would be transformed if only I could drive a car, quite forgetting how transformed it has been because I cannot." The trouble is, we get out less today, and the resulting alienation from nature is contributing to environmental problems. We are suffering from an extinction of natural experience. "I wish to make an extreme statement," said Thoreau. "Walking is about the genius for sauntering. It is not about getting somewhere, but being somewhere." Edward Abbey was blunter: "You can't see anything from a car; you've got to get out of the goddamned contraption and walk, better yet crawl, on hands and knees." By being in places, we put nature back at the centre of human affairs.

During this period of preparation, the bat came as a messenger. On a warm spring evening my daughter calls from babysitting. A bird is in the bedroom. I slip round to the neighbour's, intent on a quick release, and find instead a bat, silently gliding around the room. I stand still and watch its path of laminar flow, sweeping up over my face, and down behind, filling the room, and yet to my ears in utterly disconcerting silence. I open the window wide, but the bat seems to pick up the central spar on its sonar. For close to an hour, I try to shepherd it out with a towel. I fail, but we grow close. The light dims towards night, and still it glides and fills this mental and physical space. Then suddenly, it settles behind a yellow curtain, and I step forward, and pick it up. It's a mistake, though I don't know it yet. I carefully hold it with neck between finger and thumb, wings and body enclosed in my palm. It is a large pipistrelle, and fits my hand nicely. I intend to show the other two, for it is not often that you see a bat of such beauty up close.

Before I take a step, the evening slips gear. The bat slowly swivels its head, around what seems like a full 180 degrees, and calmly looks me in the eye. The world stops. Eye to unknowable eye. Then the pip yawns wide, and bites down hard on my thumb. No warning at all. I blink, once, twice, and grit my teeth. I open my hand, and it hangs down, teeth gripping, blood dripping, our hearts hammering. I change plan and step towards the window, put my arm out, and shake it in the humid gloaming. The bat flits away, and we are separated forever. It's left me behind to make calls to helplines and nurses, who don't believe what they are hearing, and then there are doctor's visits to be punctured with jabs. I am lucky. I don't need the full course of five. But I cannot get out of my mind the alien look on my bat's face. It offered no thanks for its rescue. The incident was a lesson in humility, and suggested that the undiscovered and wild are near by, awaiting expeditions and journeys.

There have been some great individual walkers, and I am not intending to compete. Robin Hanbury-Tenison says exploring should always change the world. Wordsworth is said to have walked the equivalent of 10 miles a day every day for fifty years. Coleridge, Hazlitt, Stevenson, Thoreau and Kierkegaard all walked and wrote, as have Chatwin, Solnit, Sebald and Macfarlane more recently. John Muir walked the Sierra Nevada to discover Yosemite, and helped set up the world's first national parks. Bill Bryson walked more than 2,000 miles of the Appalachian Trail, declaring it was "absolutely the most blissful, happy, restorative experience". Hamish Fulton completed a 1,022-mile walk, and resolved only to make art from the experience of walking. Francis Alÿs walked urban streets carrying a punctured can of paint and unravelling a knitted jumper. Richard Long walked 600 miles in twenty days on a trip from Aldeburgh to Aberystwyth and back, carrying one stone from east to west, and bringing back another. From one sea to another, never linked before in this way. By all these means, land is animated by changing space into place.

I began with one ten-day walk, and ended up walking for a year. Time distorted. I made different types of walk. The multi-day walk, where the continuous rhythm changed perceptions of light and time, but during which I did not have enough time to stop and talk. The single-day walk to link places and meet people. The layered walk or boat trip, revisiting places in different seasons to see how things change. It was a year containing forty-five days of walking 400 miles and boating another 100. I did count, but who really cares about the number of miles? What matters is what happened during this reinhabitation. Some days I was scorched by sun, others battered by wind and snow. Some days the fog and mist closed the landscape, on others the air was clear and the vault of the sky so vast I could see to another age.

I was wet and cold, dusty and sweaty, content and sad, welcomed and lonely. I walked alone, with friends, with family, and alone again. As local people say, huge skies are a special feature of the region, bringing space, air, freedom and the sense of a long land that is both near and far. All the way up the coast, there seems to be a settlement on a slight rise ahead, or a pier jutting out to sea. When you drive to them, you do not notice a few metres in height. But when you walk 5 or 6 miles, they seem to rise from the sea or saltings like great castles. Every time you look up, they seem to shine in the sunlight.

These walks changed me. Most significantly after a ten day walk. I felt I was carrying an imprint of the sun holding position somewhere slightly behind my right eye. I had gone east, north, occasionally west inland and east again, and so the light was almost always ahead or off to starboard. It left me with an imbalance, and a sense that the whole world was luminous on one side. When dark clouds raced over the water, it was slate grey and menacing. But when the sun was out, the water became a shimmering mix of silver and mercury, and I was lit from below as well as above. When the tide receded across the wide mudflats, distant container ships elevated as mirages, or sank into perfect reflections. Birds invaded the muds. After a baking walk one day along the sands, I drove west, and the light refracted from behind low clouds and created a piece of linear rainbow. I had never seen such a thing. It pointed the way home as the beach traffic streamed out of the town. This luminous light of the coast stayed behind my eye for a couple of weeks. I carried with me the vast skies, stretched lands of golden cereal, dusty combine harvesters, sea walls of dried grass, thistledown and golden samphire, and white sails gliding across the land on invisible creeks. There was also the hammering of hail on a river wall, drenching rain in a pine forest, crisp hoar-frost on grass at winter's dawn. I heard the curlew and redshank, the outpourings of skylarks, and the crump of waves on the beach.

I have the coast in front of me now in this assortment of stones, shells, badges, china, leaves, bark, bog-oak, feathers, cartridges, bones. Their textures and shapes contain larger stories, and these are part of yet more patterned aspects of land and seascapes. There's a toffee-coloured stone with a hole from the shingle piled on a bomb-testing pagoda, exquisite blue and green shards of Victorian porcelain, a corner of fisherman's cork smoothed by the waves, a translucent moon-stone half-covered with lichen, a rusty 3-inch nail from a sea wall, bark of elm scoured by larvae of beetles, a pine cone from a shoreline forest, the featherweight bleached bone of a bird. These things talk.

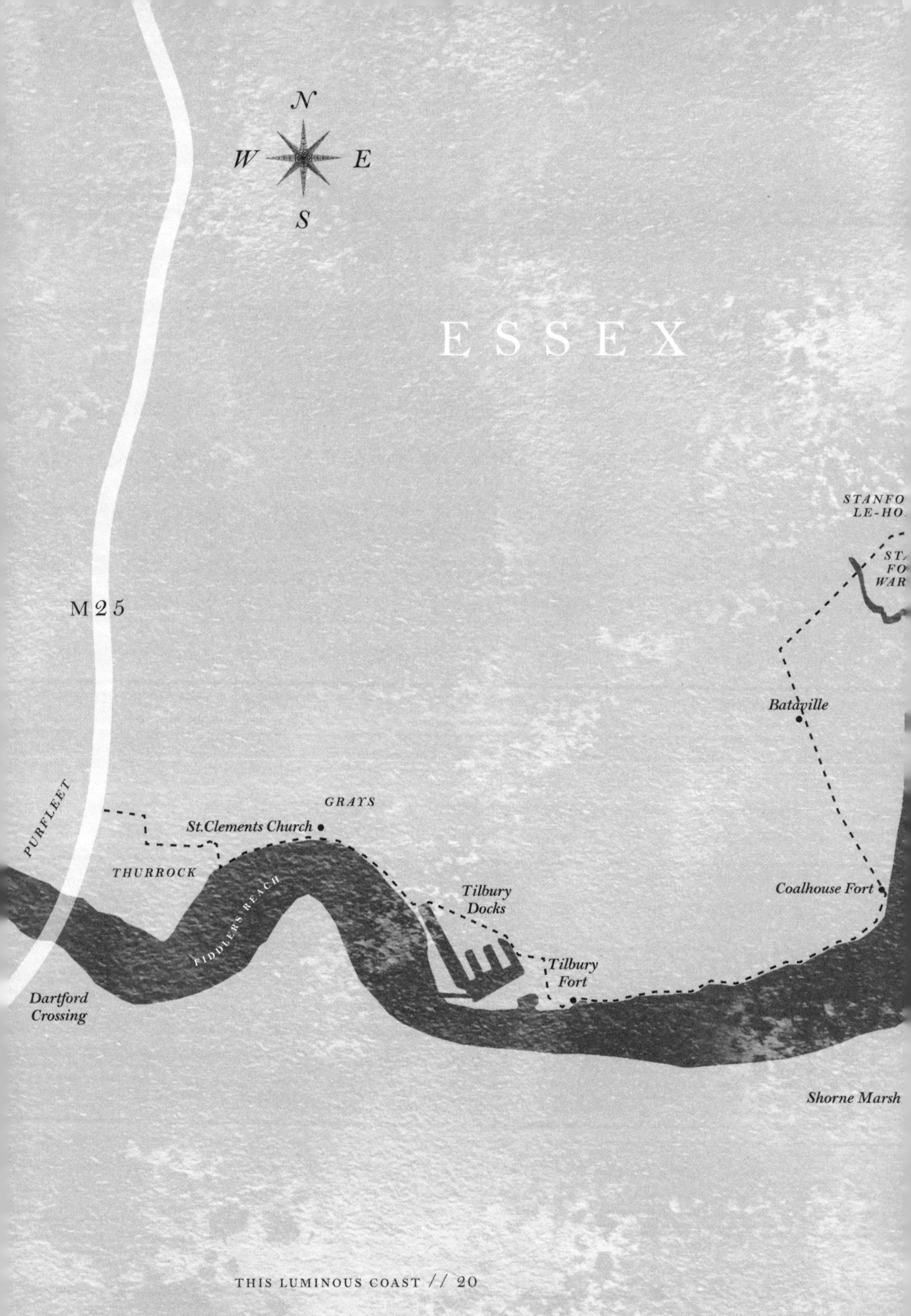
N
W
E
S
ESSEX
STANFO
LE-HO
ST
FO
WAR
M25
Batavville
PURFLEET
GRAYS
St.Clements Church
THURROCK
FIDDLERS REACH
Tilbury
Docks
Coalhouse Fort
Tilbury
Fort
Dartford
Crossing
Shorne Marsh

Chapter 1

Bowers
Marshes
BENFLEET CREEK
CANVEY ISLAND
FOBBING
arys
rch
Fobbing
Marsh
CORYTON
Shell Haven
RIVER THAMES
KENT
igham Marsh

Chapter 1
There Be Monsters

I am standing under a road that descended on the lands of Essex and Kent and their dividing river. It didn't evolve from an ancient green lane, emerge from a farm track, or follow the route of a turnpike. It was dropped from above, this orbital macadam that goes nowhere except around. I look south to the cream piers suspending this grumbling motorway that leaps into thin air, held by spreading claws of metal hawsers, then up at its curving underbelly 60 metres above, and listen to the monster roar. At the edges of ancient maps are depictions of those beasts and wildmen once thought to be a threat to the known and civilised world: chimera griffins and elegant unicorns, pachyderms and prowling beasts, and barbarians too. Monsters have loomed large in our imagination, and fear of them has helped to define us. Yet what is wild is being rolled back in the face of industrialisation and urban expansion. But a succession has occurred. Humbaba, Gorgon, Grendel and all the old crew are gone, and the new threat, especially to the modern societies that believe they will go on forever, is consumption. It's the overuse of fresh water, clean air, biodiversity, fossil fuels and metal resources that is now putting our blue-green planet at risk.

I set off walking from a scrap of grass by a white cliff, remnant of the old chalk works, dwarfed now by steel and concrete. A beer bottle hangs in an ash tree. Scattered rubbish litters the green. Cars and vans rattle by down here. This doesn't look to be a good place for walkers. *Keep clear* say the signs on the road. A lorry driver leans forward and peers at me, and thunders by. The ancient communities of Purfleet, Aveley and Ockenden reside in this long dragon's shadow, surrounded by link roads and industrial debris, their mental space permanently invaded. The first Dartford Tunnel was built in 1963, another in 1980, and the motorway itself completed six years later. It was called the Purfleet-Dartford Tunnel, but Essex lost to Kent in the name game. The QEII Bridge was opened in 1996, and this giant is nearly 3 kilometres long, contains a tidy 19,000 tonnes of steelwork, and carries

150,000 vehicles daily. The M25 has only been open for two decades but as every driver now knows, it feels full. In the sky, the vehicles rush across the curves of concrete. Everyone is on their way to somewhere else. In fact, for most drivers, these shadow communities do not exist at all, even though Purfleet achieved some notoriety by featuring in Bram Stoker's tale when Dracula bought a house here. But who is going to drive a stake through the modern monsters? Where are Gilgamesh, Perseus and Beowulf when we need them?

No one smiles. Men stand at grimy factory gates and blankly stare, coffee cups in hand; lorries growl by on roads flanked with eroded and weedy pavements; a man snores at a security point. There are office blocks with broken windows, bleached curtains flapping from long-empty rooms. Warehouses are battened down as if for a coming storm. Metal grilles are bent and twisted, paint flaking on concrete. I feel anonymous, slowed down, and trudge on under low grey skies. Isn't there going to be any sun this summer? I later read that Iain Sinclair also found everyone along this stretch avoiding eye contact in his *London Orbital*. "Don't misunderstand me," he says, "I love the place." There is noise in the background and foreground: the constant grumble of motorway traffic and clanking factories near and far. Then after tarmac, concrete and abandoned roadworks, I find the hidden path by Proctor and Gamble's detergent factory and, squeezing past an old flint wall clogged with ivy and decorated with fluorescent graffiti that says *Bile, Bile*, come to the pilgrim's church in an oasis of wildness.

St Clement's church has a circular nave in imitation of the Holy Sepulchre, and churches built on this plan were funded by the Knights Hospitallers and the Knights Templar. That austere church in Jerusalem is part functionally bleak, part kitsch, contested by different factions, and no model for anything, it seems to me. This one, though, is a simple riverside church, lately famous as the location for John Hannah's eulogy in *Four Weddings and a Funeral*. George Morgan described this church as "standing lonely in the marshes" just before the Second World War, but where are those wetlands now? The church has stripes of black flint and light-grey ragstone, and the towering factory is red and steel grey. At the entry to the graveyard is a blue board celebrating this patch as a wildlife sanctuary. Not surprising, really. There's little else that is green hereabouts. The church door is locked, but by the tower base it looks as if badgers have dug tunnels. Here are graves, too. I find a marble one for Albert Smith aged 64, died 1924, and his wife, another Hannah, Elizabeth, died 1928. *Peace, perfect peace*, it says. In 1906, a coffin discovered in a

Queen Elizabeth II Bridge

sealed vault was found to contain a body pickled in rum. It is difficult to know whether to be uplifted or entirely crushed by the surroundings.

Just beyond the church and thickets of brambles, I glimpse the grey Thames for the first time at Fiddlers Reach, where three drowned violinists play tunes from below the choppy waves. This is the line that can be walked, land to the left, sea or salty estuary to the right, and if I find a way to keep going I will eventually make my way around the entire coast. I head along the narrow concrete river wall. Down in the mud are rusting supermarket trolleys. No patch of modern wasteland is complete without one. The graffiti on the walls shows that people have been here and left their marks. They brighten the oppressive morning. Banks of fennel grow through the cracked concrete. I press on through Grays, past a bleak recreation ground where a woman walks two great leashed hounds that are marking the goalposts on a football pitch. I pass the listing Gull lightship mired in slick ooze by the boatyard, and come to the first of many new sacred places of the coast: a bright painting for *Ozone* and *Whys aka Wants.* Two graffiti artists killed eight months earlier by an underground train at Upney sidings. I stand and wonder why they chose this place by the water for the memorial. And then a grandfather is pushing a small child towards me. "Morning," he says. And smiles. That's all it takes. He's gone, but already the place looks better.

There's a strong sense of swamped history along this reach of the great river. Richard I granted the manor of Thurrock to Henry de Gray, and Grays-Thurrock was known this way for centuries. Fobbing, Corringham and Mucking are each villages with more than 1000 years of antiquity. Further back, the Northumbrian Christian, Cedd, established two monasteries on Essex soil in AD 654, one at Ithancester, now St Peter's at Bradwell, and the other mentioned by Bede as being somewhere here in West Tilbury. But interest in the specificities of place seems to have diminished. Now the new road signs by waterside housing developments indicate how little time it takes to drive to Lakeside and Bluewater, several hundred stores at each, and many tens of thousands of parking spaces. The very altars of consumption and marketed as destinations in themselves. Joseph Conrad thought this river was "one of the dark places of the earth". Now the lost places of these marshes of Essex seem to be just shadows, covered by roads and factories and retail experiences.

Beyond Grays are the great Tilbury Docks. Time to contact the world. I call Caroline of the PR department at the docks.

"I'm here," I say, and look at the 12-foot gate of vertical metal bars topped with threads of barbed wire.

"OK. I'll be there shortly," she replies, and I look again at the map to see how many miles will be saved by being allowed in this back door.

She arrives and flourishes a great key. But the lock is rusted fast. We look at each other. No one comes this way very often. She makes a call, beyond the line of duty, and then another. I look at the map again. And then a surprisingly jovial workman in oily overalls arrives, carrying a can of WD40, and clink. I am in. He waves away thanks, and slips back into the shadows. I reach for a hard hat and fluorescent top, and we walk past enormous mills and rows of silos that suck imported wheat from ships faster than any other facility in the world. Who'd have thought it? Tilbury Docks cover 180 hectares of former coastal marsh. It was from these docks that the £10 Poms left for Australia, and where West Indians arrived to monochrome England of the late 1940s and early 1950s. Today, the port is gearing up for all the extra business coming with the preparation for the east London Olympics, or as Stratford was administratively part of Essex until 1964, perhaps

Gull lightship, Grays

the west Essex Olympics. More development. More improvements, they say.

Tilbury is unkind after the silos. I say goodbye to Caroline, and promptly get lost. At first, I notice that every lorry driver smiles and waves, as if they recognise a fellow long-distance traveller. Nice. But then somehow I take a wrong turning, and find myself wandering along a peripheral road edged with another brutal metal fence curving into the distance. How did that happen? Lorries now seem to roar by, and I slowly pass vast warehouses full of boxes and tiny forklift trucks fussing like worker bees. I could go back, but don't want to do the same miles twice. My map doesn't look anything like the layout on the ground. I start to lose hope. This place now feels wild, threatening, even though I can't see any real wildlife. I'm later told that there are pigeons so full of grain that they cannot fly. Could be just a good urban myth; could be true, though. Fifty years ago, James Wentworth Day wrote that these docks were infested with foxes, as they knew this was a place to escape from the hunt. Then, an urban fox would have been a noteworthy sight. None today, though, in this urban desolation.

There's someone through the fence, standing by a lorry.

"You shouldn't be in there, you know," he states.

"Yes," I say. "I do know." And it doesn't help.

But then he points to an escape route.

"Through there pal", he says.

And not many minutes later I am walking outwards past a security guard, who jumps as I creep by. The sense of being lost simply evaporates, and here is the once grand passenger terminal, now a largely forgotten backwater alongside the modern warehouses and giant cranes. By the white clapboard World's End pub before Tilbury Fort, I fall into conversation with a couple of black guys from Leyton, father and son in search of a good place to fish. I can't help, but we walk together along the concrete river wall, talking about the river and its fish, turning to watch the foot ferry beat across the steely river to Gravesend, until finally we come upon a group of four older white fishermen.

They stop and look.

"There's nothing here, mate."

"Nuffink," another says, straight away. "You'll just lose your tackle in the seaweed."

My two companions have got neutral faces on. Dad smiles, and chats. I look in the buckets, and hell, there is indeed nothing. Maybe they're right. There are no fish. They do seem to be packing up too.

"I'll be getting on," I say. There are appointments ahead.

To the west the arch of the bridge is joined now in the skyline by port cranes and chimneys on the far side. Ahead is the first power station of the coast with its own jetties pressing into the river: Tilbury and its remnant of marshy grassland sprinkled with skinny piebald nags. In the middle of the station's concrete and brickwork is another oasis. And here's Lorna, proud manager of the environment centre here and soon to celebrate a first baby. We have a look at these new habitats she has created behind the river wall. Schoolchildren come to learn about practical conservation, which hopefully changes them as well as improves this place. Under government development plans to encourage growth along this Thames corridor, some 16,000 new houses are set to be built at East Tilbury alone, and another 34,000 at Thurrock. Where will they all fit, and how will those new residents use their local environments, if at all? It was here at the fort in 1588 that Queen Elizabeth I mounted on a white charger gave her famous speech to 12,000 troops before they set out to face the Spanish Armada. There will be no repelling of the new housing though. It's a designated growth area, after all. As we think of all those new houses, and even more clogged-up roads, we spot an ancient resident of Thameside. Slowly climbing out onto the wall from a tangle of vegetation is a giant rat as long as my arm. It turns and stares at us, fearless, unmoving. This is my place, it seems to say. Always was.

"I've never seen one like that," she shivers.

It's still watching as we climb back over the river wall.

A swirling cloud of starlings fills the air with pops and whistles, and then there is silence. I walk east from the power station. I hear a gull's clanging cry, the first audible water bird of the day, then the water lapping on the shore. I know to look carefully for Lorna has warned me. This is no normal beach. It sits by the last uncapped Victorian waste tip, and the beach consists entirely of pieces of porcelain, pottery, bottles, bricks and other remains from more than a century ago. Where these have been smashed to pieces by the river, a sand-like bank has formed. There are slivers with exquisite blue design, pieces of elegant pots for cosmetic creams or foods, and broken glass everywhere, awaiting a careless foot. Bottle diggers have scored trenches up the shore in search of rare finds. Above the beach are clumps of rare stinking goosefoot, the only such specimens in all of Essex. Half-wild by industrial. This Essex riviera. And there's no hint on the map at all.

You cannot really feel entirely wild down here on the Tilbury shore, but it soon feels far from known places. For a while, the modern monsters retreat. A yacht beats its way downriver, and I come upon the old radar station at Coalhouse Point. And here's another minor shrine, a place hallowed by grief. A bench faces the water, completely covered with children's toys, plastic flowers propped alongside. There is mystery about these new phenomena and their very public expressions of loss. I walk on, but decide to come back to this Harry's bench and the coast's strangest beach. Three months later I return. The sky miraculously clears as I reach Coalhouse Fort, originally another Tudor defence for this reach of the Thames, upgraded by General Gordon in the 1860s before he went to meet his whirling end in Khartoum. Where the path from the fort joins the foreshore I find the bench again, and investigate more carefully. It is a memorial to a 6-year-old who died at the end of August four years before. In the rose light of dawn, the metal plaque says *In loving memory of a very special boy, 23.02.97 to 29.08.03.* It is an ancient desire, this wish to leave something of ourselves on the land to mark someone's passing. But something has changed. Once the berth in the graveyard was booked. Now there is no such certainty. This disconnection from future place is something of a novelty. A similar phenomenon

Coalhouse shrine

has been the growing practice of leaving flowers at the sites of fatal car accidents on roadsides. Memorial-makers are fixing memories and leaving signs. Robert Pogue Harrison suggests there is deep anxiety at work, a desire for posthumous continuity in an increasingly placeless world.

Commemorative benches are more common too, but few acquire such status as this one. Our memorials have taken worship into the country. Sacred places are multiplying. By these means, death is not a disappearance, as David Abram observes, just a shift into the vastness of the landscape. They rarely contain more than a hint of the person's past, just dates, perhaps a statement. Was Harry from this end of linear East Tilbury, from Bataville further up, or from much further away? Did he enjoy this spot by Coalhouse Point, or was it special instead for his family? What differs from these places and a graveyard is visitors to the latter know what to expect. At the same time, cemeteries have rules and memorials on the land do not: people today seem empowered to behave out of the ordinary.

The sun rises into the blue sky of Kent and its Shorne and Higham Marshes, where on a raw afternoon Pip came upon Magwitch by the percolating, gurgling muds of a dark flat wilderness, as Dickens called it. But this morning, the industrial shore turns silver and then amber and gold, as the sun blazes out from clouds on the horizon. The river is about a mile wide here, Gravesend Reach to the west, The Lower Hope north towards Mucking Flats and the distant white airfuel tanks at Shell Haven near Canvey. At Coalhouse Point the abandoned tower has feet in the water, graffiti covering its lower brickwork and a knotted rope swings over the mudflats. Wooden teeth rise from the mud and mark lines of old defences and jetties. A top-heavy container ship from Tilbury port thrashes by, and then in the opposite direction a grey naval vessel brings a deeper visceral throbbing to the land. Normally skittish redshank are quiet this morning and probe the mud. This time I walk west towards the power station, past East Tilbury Marshes and head for Tilbury shore. Since the first visit, I've found photographs of Victorian visitors on this beach when it must have been real sand, day trippers with delicate umbrellas from the upriver city. It feels remote today, even though the nearest housing is only 1½ miles away. This is the middle of a stretch that few people have reason to visit.

On the beach there are now dozens of blue, brown and clear bottles washed out by recent high tides, originally for medicines, extracts, salves, sauces. I also find porcelain pieces, fire grates, plates, clinker from coke fires, smoothed glass, smashed vases and, strangest of all, human remains. It's a boneyard. I look carefully, and find pieces of femur, hip joint, ulna, radius and fibula. But no skulls. After the 1832 Cemeteries Act and later Burial Acts, human remains were brought from the full graveyards of London and dumped on rubbish tips like this one. Laid finally to rest in the middle of the marshes, far from civilisation, these bones were not meant to come to light. A wandering graveyard.

The tide recedes. Formation-feeding white-and-chestnut shell ducks sweep their red beaks back and forth in the mud. Now the flitting redshank are piping, wardens of the marshes, and are joined by the magic of the hesitant and then rising burbling call of curlew. Winter on the Thames. The rose sunlight now lights Tilbury's twin chimneys, and inland there is pink too on the Bata factories a couple of miles across the dykes and fields. A dark harrier silently rises out of the stands of yellow grass and teasel on the tip, a small mammal in its talons. As I walk back towards Coalhouse, I hear a rustle to my right, and see a red fox loping along, also sweeping its head from side to side. It catches my eye and stops. I stop too, then blink, and it snaps into action and forges ahead before switching left onto the path. It trots off ahead of me into the climbing sun. The Tilbury fox, after all.

Back in the summer, all was quiet in the grounds of Coalhouse fort on the summer walk. Children were playing on the close-mown grass, and families sat in groups around picnic tables. I remember sitting and eating my sandwiches, but feeling distinctly and painfully alone. Strewth. I was only half a day into the long walk, yet already felt distant from my known world. I walk up past tiny cottages in East Tilbury village towards Linford and the famous Bata factory and community. Linford was called Muckingford until the mid-1800s, when the name was changed by industrialists who wanted to make it the centre for a large development. Just like that. Change the name if you don't like it. The plan failed, but the new name stuck, and it's not the last time developers changed place names on this coast. On the right behind a field containing three giant shire horses is the first of two great sprawling waste tips, the other up at Mucking. To the left are the remnants of the finest modernist buildings of the whole region. I come upon a group of council workmen clipping hedges and cutting grass.

"You've lost your snow," shouts one man.

Everyone laughs.

"Where's your skis, mate," says another.

I smile weakly, wave, walk on. So much for walking poles far from mountains. I'll have to learn to live with these jokes. It's not going to be the last time.

Full lorries hammer by every fifteen to twenty seconds, and empty ones rattle back up the hill. Whatever the merits of these communities, they are diminished on this road to the edge, the largely hidden end of the line for our wasteful society. This is where a fifth of London's rubbish is pushed out amid swirling masses of gulls, and crushed down and down by a swarm of machines. Later, at the far end of Fobbing Marsh, I will pass Pitsea tip, made briefly famous for its abundant foxes, even though the site is for the most toxic of waste. On the way to Mucking, in a tiny back lane no doubt once quiet enough to promenade along, I have to jump into the hedge several times to let vehicles pass. Drivers wave their thanks, but on several occasions I feel lucky to escape. I find a footpath, and climb over the stile into a field shaved by half a hundred rabbits that pause, and then scuttle for burrows and ditches.

Bata factory

One of the most famed of the region's nonconformist settlements rose up in the 1930s between the villages of East Tilbury and Linford. Bataville was established by Tomas Bata, and modelled on his company town at Zlin in Czechoslovakia. Bata was a friend of Le Corbusier and an advocate of the modernist movement. He was the first to make shoes on a mass-production line, and believed in the social responsibility of business, constructing a series of magnificent buildings for the shoe factory, and workers' houses with electricity and proper plumbing. The novelty of inside bathrooms and toilets was remarkable. Bata intended to create a Utopia around his factory, but he died in a plane crash in 1933, and the settlement ended up only a quarter of the intended size. But his son went on to make sure Bataville had its own newspaper and other services including shops, tennis courts, swimming pool, cinema, school, college, and drama, camera and garden clubs. Poplars were planted to lower the water table, and in every front garden was planted a cherry tree.

Some thought it a cross between a holiday camp and a prison, others couldn't bear the cult-like corporate philosophy. One former worker observed that you were watched over, but it wasn't oppressive. Journalist John Tusa, whose father was factory manager, has observed that it was "incredibly paternalistic. Nobody acts like that today, building model estates, looking after workers for a lifetime." East Tilbury exported 4 million pairs of shoes a year at its height, and there were Czechs, Sudeten Czechs (who spoke German) and Poles working alongside Essex men and women down here on the marshes in a community that looked like it was designed at the Bauhaus. Sadly, the factory buildings are closed, and now used for the storage business. Another symptom of over-consumption: we seem to have so much stuff that we have to pay to have it stored. I stand on the road near the former cinema, turning away from the waste lorries, and think it's too bad that this effort at benign social engineering effectively failed. This was not the only workers' model in Essex. Up at Silver End, window-maker Frank Crittall built a factory and community in a self-contained estate with clinics, cinema and laundries, also in the 1930s. His claim to fame: the first employer in the world to introduce the five-day working week, beating Ford in Detroit by six months.

The map tells me to skirt south of Stanford-le-Hope and north of the old Corringham refinery to make it across the marshes to Canvey Island. After the massive Mucking tip, soon to be capped and turned into a nature reserve, I pass between the church of St John the Baptist with its avenue of mature limes and chestnuts and a farmyard littered with abandoned machinery, and come to Stanford Warren, the largest area of reedbed in Essex. It is beautiful, and almost tranquil: distant traffic noise intrudes. I sit on the grass and look across the swaying green reeds, and attend to emerging blisters that are to accompany me for the whole walk. I change shoes, and walking on, see a woman coming towards me with a giant of an Alsatian. I pause. It's slavering at the mouth, and rushes at me.

"'E won't hurt you," she laughs. "'E's only a puppy." And they say all the dangerous wild animals have gone in our modern landscapes.

Another monster rears up from the lower fields and marshes by the river: fuel storage tanks, chimneys, and networks of pipes of the Corringham air fuel refinery. On these marshes is more layered corporate history. In 1895, they were chosen by Kynoch to locate an explosives factory for the manufacture of gun cotton, cordite, nitroglycerine and cartridges. The new village of Kynochtown was built, and the Corringham light-railway linked village and factory. After the First World War, the factory was closed, and the village wiped away and replaced by Shell with their refineries for petrol and other cracked hydrocarbons. Samuel Pepys called this place Shield Haven, a variation on this name being taken up by the oil company who then

Shell Haven from Canvey

adopted a scallop-shell symbol. Then in 1923, the Cory brothers of Cardiff began to construct another refinery on the Kynoch site, and not to be outdone renamed their village Coryton. In the Second World War, many of the factories along this stretch of Thames were targeted in air attacks, and at the same time all the inland marshes were scored with ditches in the shape of an X to prevent enemy aircraft and gliders landing. Industrial activity briefly expanded in the 1950s along the shore now known as Shell Haven, with bitumen, kerosene, naphtha, detergent, crude oil, nitric acid, fertilizer and ammonia plants, and row upon row of great oil- and gas-storage tanks. A new Coryton refinery came on stream at 4 am on 31 January in 1953, the very worst of timing, as it was thoroughly flooded just a few hours later. But by the 1960s, all was in decline. Coryton village was closed, like a bankrupted business, and householders offered transfers to Basildon and Thurrock. In the 1980s, most of the river jetties were removed, and now processing activities have all but ceased apart from a rusting bitumen plant and an aviation-fuel refinery, still large enough to dominate the south-eastern skyline.

This place deserves a return visit, and so on a steely December morning, I meet the team from Dubai Ports. The chief, Simon Moore, takes me around their 1500 acres, littered with rusting pipes and concrete standing, reedbeds containing hibernating great crested newts, lone brick buildings, iron gates fencing off nothing in particular, yellow grass rippling in the biting easterly wind. Here will be, soon, another reinvention for these marshes, for DP obtained planning permission in 2007, and in the coming years will open the first new port in Britain for more than a century. Some 95 per cent of UK trade comes and goes by ship, and this new London Gateway port will berth ships close to half a kilometre long carrying 20,000 containers. They'll tower over the Thames as they inch their way along a deep channel to be dredged 85 kilometres into the North Sea. This is where the railhead originally built for the explosives factory will help, as it will take 500 lorries a day off the roads. A wetland will be created on Stanford Marshes, and the site will link to Essex Wildlife Trust's new nature park on top of those shifting mounds of Mucking rubbish. There will be jobs too, and cranes of course, twenty-four of them, 140 metres high when the booms are up, more praying mantises on this coast of wide and horizontal space.

We stand by the river leaning into the roaring wind, and look to where the imagined port will appear. I can't help wondering what would happen if we suddenly consumed far less, kept things for longer, threw fewer of them away. Simon's in shirt-sleeves, determined to keep to what he's used to in the tropics. I am wrapped up but freeze. He understands the values of place and loyalty. He's a lifelong Bury fan, and like all supporters of small football clubs knows that winning competitions is a great rarity. Hopes will always be dashed. But there will be changes here. A great port will emerge alongside 650 hectares of connected natural habitats to the south, and once again this region will be transformed.

- - - - - - - - - - - - - - - -

Back on that first day of summer walking, I had passed Thurrock football ground and St Mary's Church in Corringham with its shingled pyramidal roof and square Norman tower. Rowdy Friday afternoon drinkers argue outside the pub. I slip down an alley, head down, feeling distinctly out of place. Further on are ponds and manicured grounds behind fences. Outside, I have to push through banks of stinging nettles, past fields grazed within an inch of their lives by so many horses, eventually to find the beauty of Fobbing Marsh and its curving lines of inner and outer sea walls. On the hill is Fobbing village, its prominent castellated ragstone church modelled on Hadleigh Castle a few miles up the coast. Fobbing is famed for its fiery young men who in 1381 led the Essex contingent of the Peasants' Revolt to the Tower of London, and then half a millennium later for holding the open-air heavyweight boxing championship of England. But today, everything seems to be fenced off and private. By a patch of woodland in the long grass, I come upon a couple who start guiltily, and have no idea of a route across the marshes when I ask. They stare. Why would anyone want to go across there?

On the far side is the giant flood barrier at Fobbing Horse, like the Thames Barrier part of the post-1953 response for better coastal flood-protection. Chris Barningham of the Environment Agency has come to open it up for me, and he walks across the thistledown marshes and old sea walls, white shirt-tails flapping in the wind, to guide me in around hidden dykes. To the south, a smudge of sickly yellow gas appears around one cluster of refinery chimneys, and slowly disperses. At the far end, the flare roars into life every few moments, and flame flashes high into the grey sky.

I had at first thought that the barrier was a small device, but the map lies. Here powerful motors can drop plates into the creek from a military-style concrete structure to protect the upriver communities at Basildon from tidal surges. Chris unlocks the clanking metal doors, and we climb a spiral staircase to get to the engine room above Benfleet Creek. Oiled shafts and motors cased in red and blue lie poised, rarely used but ever ready. Outside on the concrete cliff is a peregrine's nest with views across fully 8 square miles of Fobbing, Vange, and Bowers Marshes. It's a sky to hunt in, and out here have been seen sparrowhawks, red kites, buzzards, harriers and hobbies. Chris takes me across a second flood barrier, past Pitsea dump, and then I am into the west end of Canvey Island, walking along a route across the marshes marked with red- and white-poles to guide emergency vehicles should the island's two road bridges ever be blocked.

Another designed community emerged on this former sheep and cheese-making island after it had been bought by Frederick Hester at the end of the nineteenth century. He advertised seaside plots to Londoners, and out they came. At that time, London commentator, Coulson Kernahan, said Canvey was a "lonely and out-of-the-world spot [that] could be discovered at a distance of thirty miles from London". Hester called Canvey an extra lung for London, and charged £5 for each plot of land with a Christmas turkey thrown in for free. But his business model flopped, and he went bankrupt in 1905. Amazingly Canvey itself prospered. Londoners kept coming. There was no bridge until the 1930s, and the feel of an island remains today.

I walk along a mile-long straight road from Northwick on Canvey, which is being used by two men to race a glistening blue BMW. They roar up the road, screech to a halt, spin and power-turn, and do it all again. I walk on, ignored, past roundabouts and supermarkets, and along Long Road chock full with grumbling rush-hour traffic returning from the mainland. I am looking forward to putting my feet up after 24 miles of walking.

Canvey is still an escape route for many from the city though mainlanders still seem to find it strange. Islanders seem to be able to live differently. Once there were no house numbers, and postmen had to remember both house names and family names to deliver letters. Many homes are upside down, their sitting rooms upstairs for a view over the wall to the sea. Out on the eastern marshes was a concrete barge, 250 tonnes of ferro-concrete beached on the saltings, and for six decades a secret meeting place, a place of solitude, a place to get drunk. Then the yacht club unilaterally destroyed it; there were protests, not so much about the ugly barge but more about the loss of memory and a place of youthful pasts. In artist Lucy Harrison's *Captivating Canvey*, quirky and more interesting than the original island guide of the same name, is the story of Canvey's Lourdes. In the 1970s and '80s, coachloads came to a woman's back garden after she claimed to have seen the Virgin Mary there. This is not such a strange assertion. Both Ashingdon and Rayne had medieval cults resulting in the barren and sick flocking to where Mary had apparently appeared. Other Canvey characters included poacher Arthur Reid, who always wore a long coat to hide animals, and could run a rabbit to ground, circle it, hypnotise it like a stoat, and then reach in for the capture. There was Catwoman, who pushed a pramful of cats dressed in bonnets around the streets, Captain Birdseye, who wore a sou'wester in all weathers and at whom children cruelly threw stones, a teacher called Slasher Hills, Roman George who thought he was a gladiator, and the Bird Man with trilby and grey mac peddling his personal cure for rheumatism. Do islands encourage particularly odd behaviour, or is it just that people remember them better?

I hobble along to the Oysterfleet Hotel to meet members of the Canvey Rendezvous Walking Club, Valerie and Derek Lynch, George Beecham, and Graham Stevens. Whilst a wedding celebration runs at full throttle in the bar, they tell more stories of an island whose population has expanded from 11,000 in the middle of the twentieth century to 40,000 today. In the war, the island was full of guns to protect the Thames, but it was still Canvey. "Imagine yourself for a moment here on a moon-misty night in the late spring of the war," wrote James Wentworth-Day. "Curlew call on the saltings, and the bar geese laugh their ghastly mirth out on the mud – that chilling laugh which the old fishermen tell you is the ghosts of dead sailors, jeering at those about to follow them by drowning." All four agree on Canvey's still special nature, the sky, the wildness, the community spirit, the sense of otherness from the mainland. The identity of on-island compared with off. Everyone talks of being *on* Canvey, but over on the mainland you are always *in* places.

I am still interested in monsters, and ask about Canvey's own. In autumn 1954, and then again in summer 1955, two strange corpses were washed up on the south shore near Dead Man's Bay, where bodies eventually seem to appear if there's been a drowning in this wide estuary. Some said these particular bodies had legs and toes, with prominent gills, brownish skin, the first about 2 feet in length, the second about 4 feet. But Graham was there at one of the strandings, playing down by the beach swimming pool. Not a monster, but a great angler fish with razor teeth and lure. Joe Overs, leader of an obscure local sect, took photos, but these were lost. As somehow they always are when it comes to monsters, dragons, bigfeet and the like. How these angler fish swelled in the story-telling to become Canvey's monsters, no one seems to know. It's another good island tale. After the pub, I pass gatherings of young people, cars racing up the streets, and then see that the moon has appeared. The clouds are breaking up, after weeks of rain and unseasonal chill. Maybe there will be sun.

ESSEX
Pitsea tip
Fobbing Horse
N
W
E
S
Lobster Smack Inn
Shell Haven

Chapter 2

BENFLEET CREEK
TWO TREE ISLAND
TEWKES CREEK
Sunken Marsh
Canvey Heights
CANVEY ISLAND
Dead Man's Bay
RIVER THAMES

Canvey Island seawall

Chapter 2
The Great Tide

Every East Anglian of a certain age still remembers the 1953 floods, where the sea walls were breached in over 1000 places, and more than 300 people died in the dark, as with no warning whatsoever wild gales brought roaring waters down the North Sea and crashed them over land into houses. Such extreme events were thought to be rare, but perhaps we need to ready ourselves. On 9 November in 2007, spring tides, north-easterlies and low pressure combined to raise the sea to levels that in places surpassed those reached that terrifying January night fifty-four years earlier. The sea almost broke through at Walcott in Norfolk, and on long stretches of the coast today's higher sea walls came within a few centimetres of being overtopped.

A few days after this November tide, I came to be walking around Canvey. On this Sunday morning, I cross the marshes towards the flaming oil refinery, past kids playing football on parks besieged by gasometers, clusters of wrapped parents shouting from the half-way line. Inside the walls, people and homes are alongside industry and now faded tourist glory. Outside are beaches, creeks, saltings and the estuary. I set off from the famed Lobster Smack Inn, a reference point in *Great Expectations*, dwarfed by a concrete sea wall with its walkway now level with the pub's upper storey. A century ago, on his *Marsh Country Rambles*, Herbert Tompkins stopped here to watch the boats from the pub. "The weather did not invite merriment," he wrote. "We put our backs to the wall and our legs under the heavy table; watched the flicker of the fire upon oaken beams, exchanged a few anecdotes, and smoked our pipes in peace." What more might any traveller want? Today, a nor'westerly gusting to gale force brings the rain horizontally along Hole Haven, and I have to bend into the wind to stay upright.

Until the late 1850s, there were convict hulks anchored to this bay, and local fishermen called it Botany. At night, each of the disused vessels housed 5 or 600 prisoners in pitifully crowded conditions, and by day the men, some consigned for stealing as little as a loaf of bread, formed the first recorded chain gangs of forced labour. In the neck of the same creek today, two rusting tugs are riding the angry gun-metal seas, pitching and yawing, and waves are slapping over great black buoys. Over on the far side where the Medway flows out of Kent is the broken wreck of the *USS Richard Montgomery*, sunk in 1944 and chock full of explosives. It's still a potent time bomb: naval experts say it would throw material up 10,000 feet if detonated. I walk down the jetty by the old customs watchtower, carefully over the mud-slick wood, and turn to look back at the sea wall. All around the 15 mile perimeter of the island are 5-metre concrete walls, holding in the houses, keeping out the water. Later as I walk along the inside of the wall, I will see deposits of bladderwrack and smashed portions of squid and other sealife, brought over by those recent massive tides. Near Dead Man's Bay, I pick up a smoothed cube of brick, caught in the gaoling eddies by the wall here, and then turn back by Labworth's white seaside restaurant perched over the south sea wall, Ove Arup's first commission in the 1930s which set up his collaboration with Lubetkin on that famous penguin pool at London Zoo. When you look inland from this castle wall, though, and imagine the weight of water when the sea is full, then you sense just how tenuous is life below sea level, even here in what seems like a mostly benign estuary.

For 1,000 years or more, some 330 miles of sea-defences have been constructed, repaired and raised to protect the land and people of Essex from the salt water. Each century contains its stories of breaches, inundation, labour-conscription, pitching and hurdling, hasty raising of inner counter-walls, all to maintain this unbroken walled fortress. Barking Marsh was flooded in the 1380s, Wrabness in the 1550s, and 400 feet of Dagenham breached in 1707 and couldn't be closed for thirteen years. This was followed by the "outrageous tide" of 1736 that flooded Bridgemarsh Island and opened a breach at Holliwell Point all the way to Burnham. Black Monday of November 1897 saw 35,000 inland acres flooded across Essex, a 1928 January tide reached London and drowned people in Westminster and Fulham basements, and then the 1949 1 March gales flooded another 9,000 acres. And all this time, the land had also been sinking.

Like many major natural disasters that remain fixed in personal and national psyches, nobody saw the 1953 floods coming. Even as that Saturday night unravelled, still no one could imagine the sea coming in with such devastation. It all began when a low-pressure zone deepened over the north of Scotland on Friday 30 January, producing sou'westerly winds over in the North Sea, and driving water from south

Lobster Smack Inn

to north. In the early hours of the Saturday, just one tide away from disaster, the high tide was a foot lower than it should have been at Southend. Then as the depression travelled rapidly east, the winds veered to northerly, and water from the Atlantic was suddenly driven back into the narrow North Sea. By six o'clock, air pressure had dropped to 968 mb over the Orkneys, bringing the sea level up by an extra foot. Still weather forecasters were predicting no more than a "vigorous trough of low pressure". But the pressure over the North Sea would lock between 968–976 mb until midnight on the Saturday, by which time so much water had piled up in the North Sea that the defences on the east coast would at first be threatened, and then in a short few hours overtopped, breached and then soundly beaten.

On the Friday evening, gales gave an indication of trouble to come. Constant winds of 75 mph were recorded over northern Britain, gusting to 100 mph. At Southampton, the sailing for New York of the 81,000 tonne liner, *Queen Mary*, was postponed, and my stewardess grandmother had to sit out the night in port. At Ullapool, twenty-seven herring drifters in Loch Broom broke anchor and were swept up to be stranded inland on fields, though one doughty 70-year old skipper called it no more than "an exceptional breeze". But a quarter of all the Scottish fleet would be lost that night. Much worse was to occur in the Irish Sea, where the ferry *MV Princess Victoria* left Stranraer just before eight the next morning carrying 177 passengers and crew, expecting to take three hours to cover the 36 miles to Northern Ireland. It was to founder and sink, taking down 133 people, the worst national disaster in peacetime for a quarter of a century.

Sometime ago I walked the famed Antrim Coast Road from Cushendall and at the end stopped at the memorial on the front at Larne. The victims' names are carved in marble. Liam Kelly was 8 at the time and says no one has really got over it. A service is held every year and there are still tears. Liam's president of the Historical Society, and they added the names to the memorial on the fiftieth anniversary of the disaster. His neighbour, John McKnight, is over 80 now and was cook on the *Victoria* that day. As the ferry passed Corsewall Point from protected Loch Ryan, he felt a massive thump and the whole ship shuddered. The severe nor'westerlies, combined with heavy sleet and snow, had burst open the stern doors. Sea poured into the car decks, but the scuppers were too small for the water to escape. The ferry listed to starboard, cargo shifted, and she tipped over further. The captain tried to reverse into the loch but failed, and decided to continue towards Ireland. For hours she wallowed and fought the gale alone, as rescue boats tried to find her. The ferry had no radio, but the Morse code signals continued to suggest she was still close to the Scottish coast rather than heading across the Irish Sea. It seems it had been impossible to give an accurate position because of the listing of the ship. Two lifeboats spent hours searching, a shattering task in itself. The destroyer *HMS Contest* was ordered to proceed to assist the *Victoria*, and she steamed south from Rothesay at 31 knots and reached the ferry's supposed position at noon. She too was in the wrong place.

The ferry was still alone in the mountainous seas at 1.30 p.m. when the captain reported the Irish coast was visible, and twelve minutes later they were at the entrance to Belfast Lough. But now the engine room had flooded, and the ferry listed to 45 degrees. Less than half an hour later, the captain ordered the ship to be abandoned, but few of the lifeboats would launch properly. A little after two the ferry sank. John remembers jumping into the water and climbing with other men onto the keel of the inverted ship. He thought that if the women and children survived that would be something. But then they watched in horror as a giant rogue wave bore down, flipping one lifeboat full of women and children and drowning all the occupants. Of those who survived that day, all were men. Four merchant vessels put to sea from Belfast Lough, but could not get close to the remaining lifeboats because of the fierce seas, and all they could do was provide meagre shelter until the Donaghadee lifeboat arrived. By 3.30 the *Contest* had found wreckage, and then began taking survivors on board. A lieutenant and petty officer both received George Medals for jumping into the sea to rescue a survivor clinging to a raft. Other awards were given to the skippers of the two lifeboats and the four merchant vessels and to the captain and radio operator of the *MV Princess Victoria*, both of whom were still at their posts as the ship went down. The *Victoria* would dominate the national news over the weekend, whilst unknown to the rest of the country the sea sank communities in Hunstanton, Yarmouth, Harwich, Jaywick and Canvey over on the east coast, and killed another 307 people later that same day.

In the North Sea, the problem was not the occasionally very strong gusts, but the extraordinarily high average wind speeds maintained over many hours. These prolonged and violent winds forced an extra 15 billion cubic feet of water into the North Sea, raising sea levels by an extra 2 feet. The tide came as a giant standing wave, hundreds of miles long, arriving at King's Lynn five hours before Harwich, and seven hours before Tilbury on the Thames. This night's tide would be higher than ever previously known. Now two key factors were in place – very low pressure and strong nor'westerly winds. The third was the full moon on the Sunday. Full and new moons bring spring tides, when high tides are highest. As people in coastal communities in Lincolnshire, Norfolk, Suffolk and Essex went to bed that Saturday night, none knew of the coincidence of these three circumstances. None had forward warning of the risk. No one would be able to do anything but try to save themselves and their families as the tide came in, and in. What lessons we ought to learn today from this night.

Signs of unusual sea levels were noticed that Saturday afternoon, when no ebb occurred. The wind "seemed to be holding the water", reported one policeman. Across the North Sea, where the loss of life was to be far worse in the Netherlands, no ebb was recorded either. That night, the high tide should have been 5.5 to 9.6 feet above Ordnance Datum Newlyn from the top to bottom of Essex. It was to be 13.1 feet at Harwich, 14.4 feet at St Peter's by Bradwell, 15.9 feet at Tilbury, and 16.9 feet at Barking, the highest on record. By five o'clock, as football supporters from across the region were trudging home for tea, sea water crashed through the dunes and sea walls in Lincolnshire, and local crises were unfolding at Sutton, Mablethorpe, and Saltfleet. Every automatic tide gauge in the Wash, and eventually

Labworth Restaurant

all those to Southend, would be destroyed by the enormous weight of water. River Board officers at King's Lynn became concerned when expected high water levels were passed two hours before high tide. But still the worries and actions were localised. Then the 7.27 evening train from Hunstanton to Lynn ran headlong into a wall of water a mile inland from the north-west Norfolk coast, and was to be stranded for six hours. And now the first serious disaster occurred. Forty bungalows on an estate in south Hunstanton, home mainly to American servicemen and their Norfolk wives, were flooded, and all but three collapsed and were swept away. Sixty-five people drowned.

At Sea Palling, 6 metre waves burst through a 100-yard gap in the sand dunes, and rushed inland. Yarmouth was engulfed at ten o'clock, still three hours before their high tide. At Southwold, our family friends Rene and Donald Horwood, one of only two families resident upriver on the Blyth at what is called the Harbour In, noticed the water coming up their garden path and lapping at the door. They'd never seen such a tide. And still the wall of water poured south. At sunset in Harwich, it was difficult to stand in the wind, and heavy seas were crashing deep into the harbour and Stour and Orwell Estuaries. Anxiety increased as the six o'clock BBC news bulletin reported the *MV Victoria's* loss, and local fishermen and yachtsmen shared their worries about the poor ebb and severe winds. Yet down at Tilbury, a lookout near the fort recorded only that the "wind was fresh, but the water calm as a mill pond". In the previous eight winters fifty-six warnings had been received from the Met Office about high tides associated with north-westerly winds, but none were this high. It is difficult today to imagine the lack of real-time information. No regular updates of news; no mobile phones; no internet; landlines linked only via local exchanges constrained by the need for human operators; few families with phones at all. No one in Essex knew of the Hunstanton train-stranding and drowning disaster. Today, this is inconceivable.

By ten o'clock in the evening, all down the Essex coast the sea had completely covered the saltings, and was working away at the sea walls themselves, filling rabbit burrows and badger setts, loosening cracks in the clay, undermining foundations. At Maldon, a group of fishermen decided to sleep on their boats to keep a close eye, and half an hour later all the rest had gathered on the quay quite unable to get out to their boats. One out on the water recorded gleefully, "You should have seen `em, there they were standin' up on the benches along the prom, squawkin' and hollerin' like a lot of old crows." They did, though, spend the night looking after their rivals' boats on the Blackwater. The drama was mounting at every location, ready to play out in the bitter wind and fearsome conditions. At the time no one knew where the sea would stop. Now the events of the last day of January and the first few of February were to unravel almost minute by minute.

In Harwich, the water tops the quay and starts to spill into the streets. A police constable hurries from house to house from eleven o'clock, knocking on doors, and then finds himself pursued down the street by rushing waves. By half past eleven, the sea is pouring into the town, and a 3 foot wall of water is crashing into houses and seething up roads, almost stealthily beneath the roaring gale. A fisherman on Alemeda Road is warning neighbours on his still-dry street, comes home to gather up his wife and children, but then a wave crashes through their small house. He manages to carry them one by one to higher ground. Elsewhere, families stand at the top of their stairs, shaking after explosions of water smash through windows and doors below. They wonder where the rising water will stop. Looking out, they see waves crashing into fences and pounding houses, and pigs being washed along by the current, and do not know what they will do if the water comes into the bedrooms.

Further south at Walton, just like on Southwold, Fobbing and Tilbury Marshes, the sea water sneaks in the back door and whilst everyone is looking seawards to the waves and the wind, it drowns houses from the land side. At Jaywick and Canvey, people are sleeping behind 1,000 year-old sea walls, trusting to history, but will be betrayed. The walled fortress of Essex is under siege, and soon will surrender. Down on the Dengie, the lonely 1,300-year-old St Peter's Chapel is next to go under. Near by, the cottage of the Linnet's, red brick and creosoted clapboard, sits in a tiny copse. It is on the front line, on the saltings side of the sea wall. They are the marshland's most notable wildfowling family, as well as employed by the River Board, and have seen the effects of high water before. The straight road from the chapel leads to the Cricketer's Arms, and that night's poor ebb and high winds are the topic of many conversations. Walking back at eleven o'clock, the son struggles into the wind, and then sees that the saltings are drowned and water is slopping at the walls. Then there is a clap of thunder, and the sea surges through a hole in the wall bringing a wave several feet high. He escapes just in time with his father, Walter, who's seen nothing like this is all his 80 years.

Around in the Blackwater, at Canney Farm near Steeple, a farmer watches the rising river. After the final milking of his magnificent dairy herd of seventeen prize-winning Friesians at 11.30, he comes out of the shed and in the bright moonlight (by now, the ghostly moon is beginning to light Essex, but the wind is not letting up), he looks across a white landscape. Like others from Foulness to Felixstowe, he thinks at first it is snow. But then he realises his mistake, and sees the tide rushing across the fields towards the barn. He is reluctant to turn out the cows for fear of them catching pneumonia, a fatal decision. He runs to a neighbour's to find out the time of the high tide, but when they both return, the fields are all waves, crashing onto the cowshed, some lashing right over the roof. This is not just water, it is the full-blown sea, with waves and currents and now 4 to 5 feet deep. They try to get the herd to move, but the animals fear the water too much. One by one, through the night, the cows lose feeling in their legs, and pitch into the water. The old bull, though, eventually stands for thirty-six hours in the freezing water, and survives. All the rest of the famed herd is lost.

The water strikes Jaywick in the back at a quarter to two. In her remarkable account, *The Great Tide*, Essex county archivist, Hilda Grieve, winner of the British Empire Medal for services to emergency management in the war, writes, "The Jaywick saucer filled in a matter of fifteen minutes or so, so fast that people died in their beds without moving." Some two billion gallons of water pour into Jaywick.

The bungalows and chalets do not have outside staircases like those on Canvey, so residents have to break into roof spaces from the inside. They fight short battles against the crashing icy water, grabbing children and older relatives. Rescuers move in boats from house to house, but the winds are so strong that it takes one boatman and a police sergeant twenty minutes to row 15 yards at one point. When they get to houses, one holds tight to the outside of the house, whilst the other tries to take people on board without capsizing the boat. Hour after hour they toil. "It was a dreadful task," said a Red Cross official, "we gave them hot drinks and off they would row again." Another young police constable crawls a mile along the seawall from Jaywick to the Butlins camp at Clacton to call the main police station at five to three to report that 500 people are trapped in Jaywick. With waves crashing over the sea wall, he then crawls all the way back again to help the grim rescue effort. It takes three firemen and a Salvation Army band member two hours to cross 30 yards of deep water to reach two people. Yet after the initial panic came a long wait. Said one young man, "The scene when I looked out of the roof was amazing; all I could see for miles was water with just the tops of bungalows jutting out. My father's two elderly sisters lived in a bungalow opposite us … and were eventually found still in their beds … Then there was a period of utter silence and it must have been for some hours that I stood looking out of the roof waiting for help."

The water has crashed in the back door of Southwold too, turning the marshes into a raging sea and making the town an island for what would be forty-eight hours. Rene and Don are joined at their house by fisherman Frank "Workie" Upcraft, who confidently says he's never seen the water above the skirting boards. Then they have to open the back door to let the tide through, and it keeps on rising. Two hours later, the sea wall breaks with a thump, and the low marshes between harbour and town are filled with 16 feet of water. They dash upstairs, and then realise they have to get to their neighbour's, where paralysed Mrs McCarthy sleeps downstairs. Don and Workie fight water and wind to rescue the terrified woman, and carry her up to wake her deaf husband. Rene then sees lights flashing across the waves from the town, but doesn't think to signal back. Later, a cockle boat appears, heroically rowed by Arthur Stannard, the brewery drayman, and American airforce officer Johnny Svboda. "We were so worried about you," they say, "we simply had to come across."

Both stay the night as the wind lashes the house, tense hours not knowing what will happen next, listening to the electricity shorting into the water downstairs. At one point they look out of the windows and see the wooden tea room from

Walberswick sailing regally upstream, lace curtains silver bright in the moonlight, heading for doom against the waiting bridge. By the next afternoon, members of the sailing club are able to wade along the Blackshore sea walls, inching back with water up to their waists, feeling the walls crumbling beneath their feet, carrying the two children and Mrs McCarthy to safety.

It is more than six weeks before they can return home, just before which fire brigades come from all over the county to pump out the marshes. Rene vividly remembers the wriggling, twisting, slithering eels, millions of them covering the marshes and clogging up the fire hoses. But then, forty years later, at a party in London, she and Don meet a couple who happen to mention vague Suffolk connections, and declare that a great-aunt had run Wavecrest, that tea room, and never knew what had happened. Rene says, "We're the only people left alive who can tell you."

Some miles across Canvey from the Lobster Smack, after I have got my timing badly wrong, and watched the sky turn charcoal, and then the landscape collapse to only a few yards' visibility as the rain slashed through the salty air, I again meet Lucy, Derek and Valerie, Graham, and other members of the Canvey Rendezvous Walking Club. We're here to walk around Canvey Heights, the new name for the former Newlands rubbish tip, capped and grassed in 2003, overlooking Tewkes Creek on the north side of the island. Once this was an area to be avoided; now people are being encouraged by artist Lucy's project to come to these open spaces and see the spectacular views of the Thames and mainland coasts of Essex and Kent. After walking in the now blustery winds for a while, we repair for mulled wine at Margaret Payne's tiny chalet opposite the Lynches in the Sunken Marsh, exactly where the water first broke through in 1953. They talk of that night. At the time, some 1,700 hectares of marsh had been reclaimed over the years from the sea, and three-quarters of Canvey's 11,000 or so residents lived in bungalows and chalets of mostly one to four rooms. All were enclosed by miles of feeble earthen sea walls.

The first person to notice a problem on Canvey itself is Derek's father, a River Board man who built his own house when moving to the island, and who that night goes up onto the walls overlooking Tewkes Creek just before midnight. In the hard silver moonlight, he sees a fleet of water where there should be islands across to mainland Leigh, and the sea is lapping over the wall at his feet. He rushes to wake his wife and son, and with another Board man begins knocking up as many people

Sea wall graffiti

as possible. "The tide is in," they shout. The walls here are a foot and a half lower than on the south side of the island, and this is soon to consign these northern neighbourhoods to 6 to 8 feet of bitterly cold sea water. Derek himself wakes his grandmother next door, who only has a single-storey chalet, and they come back to the bungalow, just a few metres from the sea wall. They retreat to the loft, and so far, apart from the howling wind, not a lot seems to be happening. Mother decides it's time for a cup of tea, a perfectly reasonable request in the circumstances, and Derek is standing by the gas hob downstairs watching the kettle when he hears the sound of trickling water. He turns, and sees water jetting through the key hole halfway up the back door.

Bang. The door crashes open, and the bitterly cold North Sea is inside. He scrambles across the kitchen, fighting the swirling water, and just makes it to the stairs before the house fills to the picture rail. They are trapped. It is the speed that is shocking. With the wind roaring, and icy sea inside the house, the electricity off, and nothing to drink, life seems to hang by a thread. Outside, Derek's father is down the road when the waves overtake him, and he leaps onto an iron fence and grabs a tall post, and there he stays, up to his chest in winter water all the night. Across the road, an elderly couple he has warned climb onto a wardrobe in their tiny bedroom, but later it gives way in the pitch dark, and the woman tumbles into the water and drowns. Like so many others, the husband is quite helpless, even though he is so close. All the clocks in Sunken Marsh stopped between 1.42 and 1.47 a.m. as the rising water reached past the mantelpieces.

In our walled-in fortresses of East Anglia, 1,000 years of protection count for little when finally nature takes the upper hand. Then there are but seconds and inches between terror and survival, and terror and death. Our coast is sinking, our sea is set to rise by three-quarters of a metre this century, could be less, could be more. We sit here, hoping for deliverance. We now know so much more about what climate change could bring, and yet appear to be doing far too little. Since 1953, Derek has watched from the same home as the walls have been raised, and defences strengthened. Will this be enough for the challenges likely to follow?

In all, more than 230 yards of sea wall around Sunken Marsh were breached and scoured to ground level that night, and the neighbourhood came to be described as a basin of death. Many houses had outside stairs, so that residents could live in the loft in the summer, and rent the main rooms to holidaymakers. But as the water gushed in, people had to get outside in 3 to 4 feet of water, and try to climb to safety. For one seventy-year-old man, "The pressure of water was so great that the lock wouldn't turn. I had to break it off with a hammer. I shall never forget the shock as the door flew open and the icy cold water poured in with overwhelming force."

It was in that second hour after midnight that the horror in Canvey's Sunken Marsh was worst. Hilda Grieve says, "The cool gallantry of one must speak for all," as she tells this story. One young woman with husband, two girls of 11 and 5, and a baby of 8 months, live in a bungalow beside the central wall. She wakes to the sound of rushing water, and sees the baby's cot floating by her bed. They jump up and force the front door closed against the water, and the husband, who cannot see well as his glasses are lost, holds up the baby, and then climbs out onto a windowsill. The youngest girl perches on the sewing-machine table, whilst the other stands with water up to her shoulders. The mother then sets off to try to swim the 30 metres to the central wall for help. Halfway, she realises the current is too strong and the water too cold, and returns to climb the outside stairs to the loft. She tears blankets and bedding into strips to make a rope, and leans out over her husband 8 feet below. First she pulls up the baby, then the 5-year-old, who is almost too afraid to have the makeshift rope around her armpits. Then the heavier older girl is pulled up and pushed from below, and finally the husband grasps his wife's hands and hauls himself up to safety.

But some tragedies are almost too painful to bear, even after all these years. A few hundred metres away lives a family with nine children under 16 years of age. When the water bursts in, Father climbs on a table to break a hole in the ceiling, and lifts seven of the children one by one into the roof space. Then the table collapses, and Mother is left standing in the water, holding onto the two youngest boys. During the endless, bitter and dark night, both die in her arms. There is nothing she can do. At eight o'clock in the morning, a third small boy falls through the ceiling, and an elder brother jumps in and stands in the 5 feet of water holding him up. Eventually, his legs go numb, and he has to let the small one go and drown. He hangs onto a door until the first boat eventually arrives. No accounts do justice to the lonely despair, the shouting and urging, and the clawing sense of failure as those smallest children died. All survivors will remember this night for all their lives.

On Foulness Island, the sea wall breaches in a dozen places as the waves come over the top and gouge away the soil. The view as the dawn light appears on a clear and surprisingly still day is enough to take everyone's breath away. The wind has stopped, but the water is in. On that cold Sunday morning on the first day of February in 1953, Sam Self comes downstairs in his house at the north of the island to meet water coming up the stairs. They look out and see that the land has disappeared. It is as if everyone has been secretly transported far out to sea. Only the upper storys of the houses can be seen, along with a few surviving trees. Just the church and pub at Churchend have escaped, as they had been built on the smallest of rises. Roy Ducker's been off island the previous evening with friends. As they cycle home, they come to the bridge to find waves splashing across the top. They have to turn back to Wakering and stay the night with relatives. On the television the next morning, the newscaster says, "There is no sign of life on Foulness and everyone is lost." Roy isn't to discover the welcome truth for another day. Remarkably only two people died on the island that night. Roy's father and two brothers later climb a water tower to wave saucepans in the sunlight to attract the attention of distant rescuers.

One farmer at Rugwood said, "When daylight came all we could see was one vast expanse of water and a few trees and haystacks. The wind was very strong and the weather very cold. From the bedroom windows we could see the gaps in the sea wall where the water had broken in."

Many animals are drowned, entangled in fences and wire, birds and animals are up in trees and on haystacks, rabbits and hares floating by on pieces of timber. That Sunday evening it is reported in the news that Foulness is still completely isolated. A DUKW amphibious vehicle tries to get onto the island, but becomes entangled in underwater obstructions and has to be abandoned. The crew themselves are not to be rescued until the Monday morning when Leigh fishermen eventually get onto the island. That afternoon, an aircraft flying over Foulness again reports no signs of life, but a second flight is more careful, and sees someone at an upstairs window waving, and then catches sight of glinting kitchenware.

On early Monday morning, an army expedition is launched up the Broomway on the Maplin Sands at low tide. It is made up of a convoy of lorries and DUKWs, including ninety Leigh fishermen with fifteen boats loaded on lorries, and they all come ashore at Fishermen's Head, except it isn't really ashore. It's a reversal. Outside the walls the tide is out; inside the walls the land is still full of water. The Southend lifeboat comes to Foulness Point on the Crouch, and sends in a boat inland to row to the rescue. The first boats arrive at Rugwood Farm by 8.30 on the Monday morning – manned by two fishermen from Paglesham. The farmer's children and wife are taken to Burnham, though the farmer stays to try to look after the farm. The residents of Foulness are not to be allowed back for six weeks, though by 19 March eighty families have returned. Today, Sam Self still feels the need to take food and water up to bed.

In all, fifty-eight people lost their lives on Canvey, thirty from Sunken Marsh alone. In the whole of the Second World War, 81,000 men, women and children were made homeless in Essex by enemy air attack. These floods made 21,000 homeless in one night. Many remember the millions of earthworms, drifting and swaying in the water, killed by the salt. But one famous image is of a group in a rowing boat on one of the main streets. Around the neck of a stuffed bear hangs a daubed sign saying *Bear Up. Canvey will live again.* On Canvey, the rescued people walked off the island, and some were then taken by bus to local schools and church halls. Many went by train back to London. Canvey was then closed to prevent looting. Large S's were chalked on doors after houses and shops had been searched. The RAF brought on mobile blowers to dry houses, and replacement earthworms were imported from the mainland. "The sea's triumph," wrote Hilda Grieve, "had been stunning."

Yet soon after, the people of the three counties were fighting back vigorously. "There had surely been no more extraordinary or spontaneous a mobilisation," she observed. And this is the untold part of the 1953 flood story – the rapid mustering all along the coast in response to this surprise attack. One particular problem was that the end of January had seen London and Essex in the middle of a major flu epidemic. As the sea came in, many key people in authorities were ill at home or in hospital. Voluntary organisations and services were called out: the British Legion, Territorial Army, churches, Women's Voluntary Service, Red Cross, Salvation Army, scouts and guides, YMCA, police, river boards, MPs, councils, schools, fire brigades, Army, Air Force, ambulance service, doctors, fishermen and oystermen. But as all these people were engaging in the rescue, they also had to anticipate the next high tide on the Sunday afternoon. Later, the perspective shifted when all 1,200 breaches along the whole coast, of which 839 were in Essex, had to be repaired before the next spring tides two weeks later. This was to be an extraordinary effort which, save for a couple of locations, would be successful in blocking out the sea again.

This race against time centred on getting huge numbers of sandbags to the holes in the walls. At the time of the flood, the Essex River Board had only 54,000 in stock. A 100-yard breach 5 feet high and 4 feet wide, though, requires 40,000 sandbags to fill it. There were only half a million in stock in the whole of the region, yet there were 1,200 holes to fill. By the Sunday afternoon, the catchment control office was requesting volunteers across the region to go to sandpits and help fill bags. Others helped to load them on lorries and railways cars, and they started to arrive at the coast by Tuesday. The river board asked for a million bags a day. At Colchester, 1,000 volunteers went to Shrub End, Rowhedge and Ardleigh to fill 90,000 sandbags in a day. By the next weekend, 8,000 civilians and servicemen were working on the sea walls, and over the next fortnight 8 million sandbags would be filled and laid. In due course, 2 to 3 feet were added to the sea walls. This meant that wider bases and broader terraces were needed, as sea walls are clay and many sit on top of marsh silts that do not bear heavy loads well.

The sea was repelled, and the walls along the whole of the coast raised, strengthened with concrete, and raised again. For fifty years, they did their job. Climate change will raise sea levels, and there suddenly emerges a belief that it will cost too much to protect all the coast. Some fields will go into managed retreat, and helpfully produce new saltings. But national and local government agencies decide to save money by letting some defences go unrepaired. No one who experienced the great tide or its aftermath quite believes how much has been forgotten.

SUFFOL
N
W
E
S
ESSEX
Benfleet Station
LEIGH-ON-SEA
CHALKWELL
WESTCLIFFE
SOUTHEND
THORPE BAY
SHOEBURYNESS
CANVEY ISLAND

Chapter 3

RIVER ORWELL
Shotley Peninsula
FELIXSTOWE
RIVER STOUR
HARWICH
HAMFORD WATER
WALTON ON THE NAZE
FRINTON-ON-SEA
HOLLAND-ON-SEA
CLACTON-ON-SEA
JAYWICK
SEAWICK
LEE-OVER-SANDS
RIVER COLNE
Colne Point
GUNFLEET SAND
NORTH SEA

Southend beach

Chapter 3
Down by the Sea

Home for my mother was a tiny pub run by her grandfather in Blood Alley in the East End. Like all city people they dreamed of having open space and freedom, and took day trips to Walton out on the Essex coast. Three adults, a teenager, and three children squeezed into her Uncle John's modest black car, and then spent hours rattling along twisting roads and tracks to make it to the sands and crumbling cliffs. These escapes encouraged the family to dream of having their own piece of land, and a plot was bought out at Abridge, and from then they went no further. Here was their green space and escape from the city. They stayed first in a tent, then gypsy caravan, next a pavilion, and finally a bungalow obtained from the British Empire Exhibition at Wembley, its asbestos walls and roof timbers blackened by fire. Eventually, the extended family moved en masse to the village. Such family diasporas are common across this region, and whole new places inland and on the coast have risen up to accommodate them.

What is it about the seaside that is such a lure? So many of us enjoy days out at the beach, some decide to stay for the rest of their lives. Everyone today recognises these desires, yet they are a recent cultural phenomenon. Before the Victorians, no one would have thought of sitting on a beach, doing nothing under the hot sun, getting in the cold water to swim. Just like mountains, which had formerly been dark places to avoid until the Romantics wrote of release and the challenges that came with going upwards, the beach emerged as a place to which urban and inland people could escape outwards. Now no longer places populated just by fishermen and their families, seaside towns and villages attracted investors in infrastructure and buildings who fundamentally changed the nature of large parts of the coast.

First came the construction of piers long enough for London's day-trip paddle steamers to berth during all phases of the tide. These were built at Southend, Clacton, Walton, Southwold, Lowestoft (two piers), Great Yarmouth (also two), and Cromer. Southend's was the giant, longer than Ben Nevis is high. The first short wooden pier, built in 1830, was extended eight years later to 7,000 feet, at the time the longest in the world. In 1877 it was replaced by the shorter 1¼-mile iron pier that just survives today. Four modern bank holidays were introduced by Parliament in 1871, Easter, Whit, the first Monday of August, and Boxing Day, and the idea of a day, week or even fortnight pilgrimage to the seaside began to gather momentum. Two health reasons were also advocated – the benefits of salt-water bathing, and the ozone exuding from the mud. New towns had to sell themselves hard to recruit visitors. When Southend's low death rate of eleven per 1,000 population was recorded in 1905, the details of these sunshine statistics were immediately sent to the London press.

By 1901 Southend's population had grown tenfold from a couple of thousand in its original village of Prittlewell, by which time many called the town Whitechapel-by-the-sea. The population continued to grow, numbering 160,000 today. People moved, and stayed.

All sorts of new entertainments were invented. At Southend, now a 7-mile frontage of buildings from Leigh to Shoebury, there were ballrooms, a theatre, restaurants, cycle-tracks, cricket and football pitches, gardens, cliff walks, circuses and menageries, water chutes on the pier, an electric stairway on the cliffs, and roller skating.

But it was not to be so easy to get in the water, for sea bathing was strictly controlled. Until the end of the nineteenth century, resorts were required to have separate beaches for men and for women. Bexhill was the first to permit mixed bathing in 1901. But unrestricted bathing would not be allowed anywhere until the late 1930s. Visitors had to use bathing machines, and those who did not want to pay were prosecuted. Local people had to fight to be permitted to bathe, let alone swim, in their own seas. Machines came to be replaced with beach huts, today so desirable in every seaside town or resort, but the "no hut, no bathe" rule continued until the Second World War. Bathing costumes themselves became a focus for conflict. From before the First World War, there had been clashes between the traditionalists and those who wanted to wear less. But even then men were not to bare their chests until the 1930s.

It was in the 1880s that steam navigation opened up the east coast at a time when road routes were horribly slow. There were not the problems in the east with Sabbath breaking as occurred in Scotland or on England's south coast, and the Belle steamers could leave London at nine in the morning, arrive at piers by early afternoon, and be back in the capital by mid-evening. On peak holiday weekends, the boats would

Southend pier

be packed, and even some of the resorts themselves became full, with visitors often unable to find overnight accommodation. The most profitable route was the husbands' boat, when London men finishing work at Saturday midday travelled out to join their families for the rest of the weekend. They carried their weekly wage packets, some of which would be drunk away on the boat as it clanked its way to the resort. There was one terrible disaster in the early days, when in 1878 the *Princess Alice* collided with a collier near Beckton gas works on the Thames. It went straight to the bottom, and 700 people were drowned, many standing up but unable to escape.

After 1914, many steamers were used for mine clearance and as troop carriers. A few had good seasons in the 1920s, but most were scrapped. Road and rail would now be in the ascendancy. Some new seaside resorts emerged, others were created as new by entrepreneurs. At Canvey, Frederick Hester bought up grazing marshes, and modern Clacton was founded by Peter Schuyler Bruff, a civil engineer and railway manager, and known as the east coast's Brunel. The pier was opened in 1871 and later extended to 360 metres in length. Clacton has grown from a population of 7,000 in 1910 to 53,000 today. Some of the roads between the pier and station still recall some of the directors of Bruff's company: Agate, Ellis, Hayes, Penfold and Jackson. Just south of Clacton is Jaywick, bought in 1928 by Frank Steadman who dreamt of a new model village by the golden sands. By 1931, 2,000 chalets had been built behind the sea wall, supposedly only for summer and daytime occupation, although many people then began to live in them year-round, leading to a long dispute with local planners who did not permit Jaywick to have running water or sewerage until the 1980s. By contrast, Frinton was purchased by Bruff with the development of an exclusive seaside town in mind. Around on the north-west Norfolk coast, new Hunstanton was developed from the 1840s by local landowner Henry le Strange, as the one east-coast resort that faces west. Later, when new settlements were prevented by stricter planning laws, caravan parks were slotted in along the whole coast, and new migrants continued to arrive. These phases of internal migrations changed the social character of the region, and created many different senses of what it was to be an insider or outsider, whether daytripper or permanent mover, bringing the accents and values of town and city to the seaside.

The plotlands of Laindon and Basildon would similarly become celebrated weekend escapes for many city families. In her family history, Deanna Walker describes a forgotten world. Today they have mostly been paved over for housing estates, and people fly overseas for their sunshine rather than seek a patch of their own land. The plotland movement began in the 1920s and 1930s, and expanded again after the war. It was a simple life: a hut on land 20 feet by 150 feet. New owners named their basic bungalows Shangri-la, The Haven and Sans Souci. These weekend retreats changed the pathways of whole families. Once one part of an extended family discovered the plotlands, then the rest tended to follow. In Laindon, there were nearly 3,000 plots, and Walker says theirs had a central place in her family's life. Coming out of Dagenham, she describes the mystery of looking through the tiny window at the moonlight and the rabbits grouped on the grass. This life meant no electricity or fridge, gas lamps, spider-infested outside toilets, the chores of road mending and grass cutting. There was always grass to cut, but also deckchairs to sunbathe in. In the evening there were walks to the pub or fish and chip shop, or trips to gather blackberries and sloes, and cooking around open fires. And most importantly, a sense of freedom, uniquely both personal and political, everyone helping each other out. There were conflicts with local people over drinking water, road access, and scrumped apples, and the incomers were called Bolsheviks for bringing in new ideas. But by the 1970s, they were suffering from burglaries during the week, and the end was near. Now these memories are all that remain in a reinvented landscape covered by housing and supermarkets.

Each year 400 million day-visits are made to the country's beaches, cliffs and marshes. On the second day of my long walk, I meet Ken Worpole at Benfleet railway station, just off Canvey Island. The day begins with cooked breakfast at the hotel, really just an extended suburban house with swirling patterns on walls and ceilings and chintzy decorations, at which a group of men sit singly at tables and shout odds across the room for the day's football matches. Ken and I intend to walk along Benfleet Creek to the north shore of the Thames and then to the point where the land turns north at Shoebury towards Foulness. He lives in Hackney now, but grew up on Canvey and then in Hadleigh next to a Peculiar's chapel. We walk along the creek in the early-morning blue haze, with sluggish seagulls in the foreground, and the marble white containers of the refinery on the horizon. The sun-soaked sea wall guides us past permanently moored barges, one with garden designed around a rusting ship's anchor, and towards the nature reserve on Two Tree Island. On this early stretch we are overtaken by runners, puffing in the warm air.

The land to the left sharply rises to about 65 metres, and at the top are the ruins of Hadleigh Castle. At the time of construction for Edward III in the 1360s the sea lapped at the bottom of these slopes. Five hundred years later, it was where the last wizard of Essex lived, and where great fairs were held. People came from far for the Hadleigh gooseberry pies and to see the Fat Lady, Living Skeleton, Fire-eater of Madagascar, and the two mermaids. Further on, we look up at the lands on which Hadleigh Farm Colony is perched, known locally as Mount Zion. It's another retreat, an escape route, bought in 1891 by the Salvation Army's General Booth just after he had written *In Darkest England and the Way Out.* He wanted a way out for London's poor, destitute and alcoholics, and also a place to train colonists for overseas. By 1912, 7,000 men had been trained, and over the years land was sold, requisitioned and rebought. Today there are 350 hectares of arable and grassland, a tea room with views to the Kentish Downs, and an employment training centre. In 2012, it'll be the site for the mountain-biking races in the Olympics. It's the nearest thing to a mountain in the whole region.

Two Tree Island was reclaimed from the sea in the 1700s for rough grazing, and originally called Oxfleet at the Southend end and Haughness at the Hadleigh end. The two trees were great elms brought down by a storm in the early 1960s. The wardens, Mark Bridges and Marilyn Pritchard, show us around the Essex Wildlife Trust reserve on the eastern half of the island. Ken remembers this as a former waste tip, and here is another example of new identities created on the coast, this time green space created over rubbish. The eastern half of the island is now a mix of reedbed, lagoon, scrub, grassland and some of the best salt marsh and mudflats in the Thames. These Leigh Marsh muds support dense beds of two species of *Zostera* eelgrass, which with invertebrates in the mud provides food for thousands of waders and geese. There are now avocets on the island, but egg thieves are still such a threat that volunteers have to protect them until fledged. In the lagoons are water voles, and in the neighbouring trees up to 100 little egrets roosting in summer, increasing in number since first arriving in the county in the 1990s. We stop at the former sewage ponds enclosed by concrete and once full of eels, and chat with a man who tells us and his son about fishing for those very eels. On the way back, we stop and talk to a mother and her three children collecting blackberries in a metal colander. So rare now, we say, children gathering wild produce. "All good things", once observed Thoreau, "are wild and free."

In the past, children naturally took to the outdoors, but these days such freedom only seems to be enjoyed at the seaside. The bond between the young and the natural world is breaking. Unstructured play is being replaced with structured experience. But playing is not a way to kill time, it is the real thing. How many adults today remember being told by their parents not to go somewhere, which they promptly did? If it was that dangerous, then the children discovered something important about the real world. Sitting at home will not do this, and neither will walking in a long line on an organised trip to the woods, perhaps in hushed reverence, as Robert Finch despaired. Yet it is through interactions with nature that we create our worlds, producing experiences and memories that may stay with us forever.

No one captures the outdoor ethic better than Arthur Ransome at the beginning of *Swallows and Amazons*, published in 1930, the first in a sequence of novels for children. Ransome was both a sailor and a spy, and married Trotsky's secretary, Evgenia. He lived in the Lake District, and this first adventure opens with the young children wanting to spend their summer sailing around Lake Coniston. Their father's in India, and he gives permission by telegram that trusts in their common sense: "BETTER DROWNED THAN DUFFERS IF NOT DUFFERS WONT DROWN."

We reflect on what seems to have been forgotten as we walk towards the buzzing model aeroplanes over the island's country park, adults at play, and then leave Mark and Marilyn to their maintenance duties. It's a lot of work, keeping a reserve going. Ken and I head for ship-building Leigh and its former deep water harbour. Today, it is silted up and famed for cockles, winkles, mussels and shrimps. Under the arc of the new bridge are traditional cockle sheds, and Ken is tempted by a tub of pink shellfish sprinkled with vinegar. Not for me, though I peer in to see if I am missing something. Leigh used to supply London by shrimp cart that travelled the 40 miles to the city every evening. This summer has been cold and wet. It's the Saturday of the bank holiday weekend, and this is the first sun for weeks. Everyone is out to make the best of it. Outside the pub lads with shirts off cluster around cold lagers. Along the sun-beaten front Leigh becomes Chalkwell and Westcliff and then grades into Southend. There's a mix of every architectural style, from white art deco, curving walls and balconies, to dismal flats of a modern marine vernacular with blue porthole windows.

In the cool shade of one of the Westcliffe Arches cafés we order whitebait and lemon. Our cold drinks are absorbed rather than drunk. We suck in oxygen, and watch the promenade of colour from the deep shade. Then we are off and into Southend proper, and Ken's memories join our wanderings along a choked promenade. Vehicles edge along in search of a parking space, and baked families mutiny. Every kind of social group jostles for space. At the end of the pier by the adventure park is chaos. Colour and bustle, packs of teenagers, mothers struggling with buggies and bags, others standing and looking and wondering what to do next. At its peak in 1949, the pier had 7 million visitors. There are fewer now, but the sands near the pier are densely packed, even though the water has retreated more than a mile out. Across the Thames is the giant Isle of Grain chimney, 800 feet high, built to take pollution far away from these shores. That chimney will hover on the horizon for days to come.

There is something wonderful about a beach. It is a place for endless creativity. The tides ceaselessly change the whole environment. Dry becomes wet, and wet then dry. Children build sandcastles near the incoming water and then watch their gradual loss. Others write their names in the sand, also to disappear under advancing water. The tide then retreats and again leaves fresh sand, mud or shingle on which to play. Perhaps, too, having chosen to be at the beach, everyone accepts the limited options for what they are, and gets on with swimming, digging, watching, drinking,

Worshippers, Southend

eating, reading or sleeping. Life is more elemental, and this is a release. You become encrusted with salt, sunburnt, hair at all angles, sand between your toes and the unique light of the sea lodged in your eyes.

Near Southend pier is a group of about seventy African men and women in full length white costumes sitting in the heat facing the muds and sea. They are attentive, listening to a preacher waving his arms, and this against a backdrop of silhouetted wandering figures on the mudflats and the spidery legs of the unlucky pier. There is still enough social space in these liminal zones for nonconformism, even on a chaotic bank holiday weekend. It was across these sands and muds that singer Billy Bragg walked, having travelled the A13 migration route to the seaside with his family. "Sometimes, if the tide was out, my little brother and I would walk almost to Holland, it seemed," he writes. The pier was known as HMS Leigh during the Second World War, and 85,000 vessels came and went from the pier head. But then one disaster followed another: a fire in 1959 stranded 500 people, another fire caught in that hot summer of 1976, and then came the destruction of the bowling alley in 1977, again by fire; then the temporary closure in 1980, the crash of a boat in 1985 that opened up a 25-metre gap, another fire in 1995, and yet another one in October 2005, so intense that the railway tracks were buckled.

After the bleak and abandoned British Gas building, we come to Thorpe Bay and sit in the shade of beach huts to escape some of the heat. Ken strolls off to a pool to wet his feet, while I lie back, unable to move for a while. The noise of the seafront slowly dissipates, and a welcome kind of silence descends, which then curiously brings a sense of loneliness. We call a taxi. "Isn't that cheating?" texts back my son.

But we need one to get back to the 1950s. We are going to finish this day on remote Foulness in a village of clapboard houses on an island still owned by the military. The contrast with the jostling seafront is immense. We sign in at the security point, and head up the spinal road to the George and Dragon. Fred Farenden, the burly landlord, makes us welcome. We are the sole visitors. Mosquitoes join us as we sit by the walled garden to enjoy a welcome pint and then our evening meal. We pop next door to the now closed church, too much money required for roof repairs, and break thick cobwebs already crossing the aisle. The rood screen is delicately painted with designs of grapes and wheat. We close the door with a creak of finality, and walk in the dying sunlight as graves cast long shadows across the churchyard. James Farr, aged 82, *entered into rest* in 1923, and his grave is inscribed with *Now the Labourer's Pain is O'er.* We spend the evening talking to locals about the tough life on this island, and in the night sleep with windows wide open, hoping for a touch of breeze. A tawny owl hoots to make the accompanying silence even older and deeper.

A resort's character is also revealed in the emptier days of winter when the front is reclaimed by local people. I come back to Southend in late winter. Down the deserted high street lies the line of the sea and the land of Kent beyond. I walk out onto the balcony of the new glass lift by the pier, and look at the water. For a while, the sky is sapphire blue and the high water an alluring aquamarine that speaks more of tropical than temperate climes. Canvey has diminished to a smudge of green, but the refinery and its white tanks are bright. This is a distorting perspective, for up here on the cliff the pier seems nothing like a mile long, yet when I get down close, it does seem to stretch all the way to the horizon.

Dark clouds race in shortly after setting off towards Leigh, and vertical rays of sunlight seem to dart up from moulten patches of water. The prom is populated by elderly couples making slow and measured progress on walks completed no doubt hundreds of times, and runners and cyclists in ones and twos, up the prom and back. The only other signs of life are men working from white vans by cafés, hot-dog stands, doughnut stalls and arcades, all making preparations for the coming Easter weekend. Groups of painters and electricians stand by ladders and delivery men carry cardboard boxes into souvenir stalls. The economy supported by the tourist pound. Perched over the sea wall is a green- and white-wooden café. *Breckfast (sic) All Day,* it says, and a huge man in tight red T-shirt sits alone and stares sullenly out. The café's stilts are in the water, one of that brave breed of building defying the effects of salt and gravity. No customers, though, even for breakfast. Elsewhere are sugared doughnuts for 70p each, no takers today; rows of slot machines, but no one to feed them; and Adventure Land, all the rides poised to go, brightly painted, greased, waiting for spring too.

Already my earlier walks have become memories that are anchors, and I find myself measuring today's experiences against them. Here is the casino, formerly Westcliffe Pool, breaching the sea wall and hanging over the water too. Along the pavement stand holly trees covered in tree lights. I sneak around the side entrance, walk out to the back balcony, and look out to sea. Inland, Southend's cliff gardens are closed because of erosion. Not much repair work going on here, too low a priority for the council it seems. Then to Westcliffe's cafés by the Arches, where we stopped, but was it the Captain's Table or the Jolly Roger? I sit on some concrete steps to the beach, and look out to the estuary again. Waves lap on the sand. Bladderwrack is piled in heaps, along with sea lettuce, and cockle and oyster shells are in abundance. The estuary water is riffled now by a breeze, but strangely it is still as a mirror over one of the rectangular sea-bathing pools submerged beneath the tide.

The magnificently modernist Argyll House is perched up on the cliff. To its left and right, though, are 1960s blocks dominating the skyline. I walk back to the pier and beyond, past a short stretch of tattoo parlours, arcades, New York-New York, Monte Carlo, all echoes of other places and none of here. Escapes. The trick is to mimic somewhere else. Next comes the Kursaal bowling alley and blue angular roofs of the Sea Life Centre, and in the distance the boarded-up gas block on the front. At some point in the 1960s, planners and councillors did great damage to Southend with prestressed concrete. They let in the now decaying Prudential building, multi-storey car parks, other half-abandoned office blocks, and the fabric of the town was ripped away. There are elegant buildings if you care to look. One is the Palace Hotel on the front, currently smothered by scaffolding and green

Southend mud flats and Isle of Grain chimney

netting. Philip Tolhurst's a local solicitor with his office in distinguished buildings near an old church, and his father was a solicitor too and more interestingly for a while manager of the Palace. In the war, he was appointed as an officer in both the Navy and Army because of his expertise in food requisitioning. The Palace was the Navy's centre of operations covering the coast from the Wash to Portsmouth, partly because Southend would so often disappear into the mists and sulphurous smogs that then so affected the Thames and London. But Spitfire and Hurricane pilots loved the pier as it acted as a giant arrow pointing to their inland airfields.

The north Essex beach coast stretches from Lee-over-Sands by the mouth of the Colne to Harwich on the next estuary to the north. On the summer walk and a few days after Southend I meet Rachel on this grey and blustery morning by piles of concrete and rubble waste that dates to the 1953 floods. The half-dozen houses down here stand on brick piers, some with wooden boarding nailed over the windows. The residents must always have their eyes on the sea wall and what lies beyond. She points to the one she once lived in, enjoying the remoteness, and I talk about neighbouring Colne Point nature reserve, a shingle spit where I spent two weeks as a volunteer warden living in a hut some twenty-eight summers before, when the sun shone every day.

We are talking about the birds that migrate across this promontory when out of the St Osyth Marsh appears a rare Montagu's harrier, its grey body and wings flashing across our path. The marsh ends and we're into the beginning of a resort strip that stretches all the way to Walton. First come permanent caravans at Bel-Air, one with tall flagpole and the statement flag of St George's snapping in the wind. After Seawick is Jaywick's Broadlands and Grasslands, a community of divergent bungalows, the short roads unmade and huddled behind a tall concrete sea wall hiding a striking beach protected by a giant breakwater of black rocks. Jaywick is one of the many places along this coast where Roger Deakin swam, the water here "icy but wonderful", he wrote in *Waterlog*. People love to live here, they especially comment on the light and sense of freedom, and also the strength of community and neighbourliness, called knock-for-knock. But the physical infrastructure is desperate, and gives visitors the impression that this is an unfriendly and forgotten place. We walk on the high concrete sea wall past a history of British cars in avenue names: Riley, Austin, Bentley, Morris, Hillman, Sunbeam and Wolseley. Rachel collects her bike to cycle back, and later tells of meeting a group of young boys excited by a catch of slow-worms.

The sun appears as I continue north past a Martello tower where once was Butlin's holiday camp, now a golf course with *Keep out* signs, and on towards Clacton, home of Clactonian man who 400,000 years ago used the earliest flint hooks and yew spears. Near the pier is the faded wall-painting of the Clacton Belle steam-train, ignored by most. I find myself looking longingly at people in deckchairs and thinking they do not know how lucky they are. It's a growing recognition of how days of continuous walking have changed the way I see the world. The emotional intensity has grown. The ups are higher, the downs lower. Along the front, whole families are gathered in front of personalised beach huts, some wrapped in blankets and with Thermos flasks at hand, others have wine glasses dripping with condensation. Down on the narrow strip of beach left by the retreating tide are children continuing their play. One boy shouts, "I'm off to save the whales." And races off. Others dig holes, throw stones at the sea, jump in the water, run around, lie down, call to Mum and Dad, Look, look. Who gaze up from newspapers or cup of coffee, smile, wave and carry on.

There is a darker side to some sea play. Two men recently jumped off Clacton pier at night, one died and the other ended up in intensive care. Tombstoning has become a craze, seeming to capture a kind of underlying desperation for something. Young men, it is mainly men it seems, have taken to leaping from cliffs and piers into the sea. After a certain height, though, jumping into water becomes like falling onto concrete. A few days ago, a man leapt off a cliff in Dorset, sank straight to the bottom and had to be rescued by onlookers. The sea is quiet today, and none of this seems possible. The pier is not so far from the beach, and the water at most 3 metres deep. But they jump after a few drinks, and the tides off these coasts have an endless capacity for surprise. I have occasionally had to swim at full exertion in the North Sea and found myself just able to maintain position against a current. The men's Olympic gold medallist at 50 metres freestyle swims at just over 5½ mph; and rip tides off some river mouths and promontories can reach 7 or more knots. "A comfortable pace for a bicycle," suggests Jonathan Raban in *Coasting*, "is a wild and dangerous speed for a body of water." It does not take much to panic and try to swim against a tide rather than use it.

Between Clacton and Holland-on-Sea, the promenade is crumbling and in desperate need of repair. At Frinton-on-Sea everyone smiles and exchanges greetings. It seems a friendly part of the coast. My guard is down: two different men call out about lost snow and skis. It's only the men, though; the wives twist their mouths in apology. I continue into retirement land. Some people move to the coast for a final escape, their last move in life. Perhaps they search for a special memory from their pasts or simply a dream. But retirement migrations are easily doomed. Friends or family might move together, but then over time some die, and those who remain feel deserted. Frinton is one such retirement community. There are many long-standing residents with local histories and memories; others pitch up and come to feel stranded on the high water line. Frinton is also notable for its railway gates. Living inside or outside them still determines social position. There is only one way in, and that's over the railway at traditional gates controlled by a crossing keeper. Inside the gates used to mean no fish and chip shops and certainly no pubs. Hotels did serve drinks, but only to local club members: a clever way to exclude visitors. The gates are identity formers, and those inside feel fiercely defensive. They are also a form of control. There is now a "leave our gates alone" campaign, as the rail company wants them automated. Not many places get agitated over railway crossings. Here they do.

Walton pier is Britain's second longest. *Loads of fun for everyone*, says the pier sign. I turn inland missing the fossil-bearing cliffs north of the town that once were famous for their deposits of copperas. Nodules of iron bisulphate were gathered from the beach by women and children, then mixed with scrap iron to form green vitriol which was used as a dye for ink production. But the ground became so polluted that nothing grew on parts of the foreshore for a century. I pass on through streets of this once thriving resort where the surviving businesses mainly seem to be furniture stores and chip shops, and head for Titchmarsh Marina in the Backwaters, known on maps as Hamford Water. On walks the quality and regularity of greetings tell me something about the place. The worst reactions, curiously, are in boatyards. Here at Walton's sailing club I try to get along a section of the seawall to the marina, but am forced back by two men: "Nah, mate, it's all private, you gotta go back."

I try to explain, and they still shake their heads.

"Forget it," one says, still grim-faced.

I give up and turn around.

I switch inland to the dizzy heights of 20 metres above sea level, turn north into a lane, and the remarkable Backwaters open out in front of me. They are 25 square miles of water and saltings made famous by Arthur Ransome's *Secret Water*, his only book set in north Essex, where "the land seemed hardly above the level of the sea, just a long low line above the water". For this adventure, the family has left the Lakes and is based on the Orwell, and Father takes the five children to what is Horsey Island and gives them a blank map. The five swallows are to camp ashore, says the narrator. They are going to do real exploring and make their own maps of these unknown waters and islands. "It's a very secret place," says the eldest, Roger. "Secret Water," say Titty, "let's call it that." And so they add their own names to the map, giving places new identities. They meet the Mastodon, a boy with sandy hair, bright-blue eyes and a face burnt brick-red by the sun. He lives on an old barge by what is really Skipper's Island, walks on splatches (wooden platforms with a rope like snowshoes) on the muds, and so leaves huge footprints. But the story has an air of a fairy tale. What parents in today's world would be sensible enough to leave young children unsupervised, including a 6-year old, to fend entirely for themselves, where the highest tides bring in so much water, and the lows reveal such treacherous muds?

Farm outhouses are topped with cracked asbestos and swarming with brambles and nettles. Ahead I can see a cluster of masts, and then come into another marina. Leon Woodrow, the warden of all these waters, is in oily jeans and T-shirt, leaning against his van, and smiling. Face lined by storm and sun, as Arthur Patterson has described watermen up at Breydon. We putter through the Backwaters, cups of tea in hand, Leon with roll-up and observant eye, a parcel of history at his fingertips. One minute, I'm an outsider walker, and the next we're afloat and exchanging salutes and greetings with every boat we pass. There are smiles too, boat person to boat person. We pass islands and creeks, then the tip of Stone Point and forge into the North Sea, up past wildfowlers' marshes, Dovercourt and then around into the twin mouths of the Stour and Orwell Estuaries. Old Harwich is on the south shore, and the container port of Felixstowe to the north. Ships with thousands of matchbox-size containers, it seems, are fussed over by great blue cranes coiled and ready to strike. We swing south into the port, dock, and then I am walking again, past the new and destroyer-grey Trinity House building. Henry VIII granted a charter to a fraternity of mariners called the Guild of Holy Trinity in 1514, which later became the Corporation of Trinity House, and in 1836 it was granted authority to take responsibility for all lighthouses in England and Wales. It's been the saviour of seafarers ever since.

The town's run down, the primary school boarded up and gates swinging on rusted gates, but Dovercourt's worse. As at Walton, with little to entice the visitor beyond pubs and cafés, the problem for seaside towns is apparent. They grew on the success of attracting domestic holidaymakers, but now too few people visit, and those who do spend little. The seaside has turned sour. Most resorts are now characterised by faded glory and peeling paint. They have plans for regeneration. They always do. But rarely do these plans build on the distinctive assets of particular places. Instead, planners and civic leaders pin hopes on striking it lucky with an external pot of gold. A northern resort was bitterly disappointed this same summer when it was not awarded a supercasino by the government. But could that really have rescued a whole community? I stay the night in the Italianate Tower Hotel, my room looking out on the spectacular harbour and distant Shotley Peninsula. The dining room is red and gold and has low crystalline chandeliers, flowers on the tables, one other sole diner and a group of six elderly people on a special night out. Between us we are hardly going to eat this town into financial success. I eat my dinner and write up my notes.

"How was the onion soup?"

"Lovely, thanks."

N
W
E
S
ESSEX
GOLDHAN
MALDON
Heybridge basin
Northey Island
Osea Is
MUNDON
Bridgemarsh Island
CANEWDON

Chapter 4

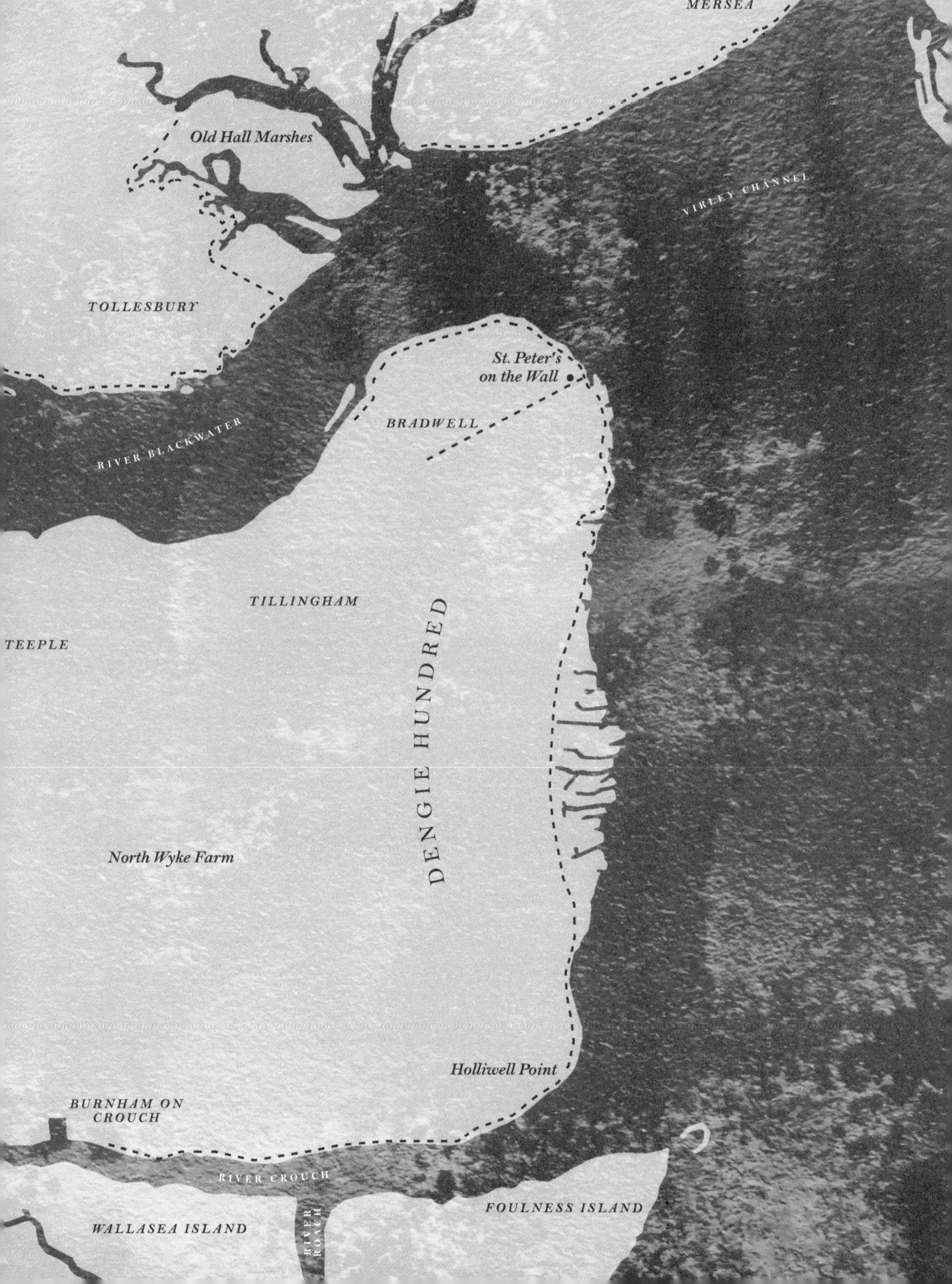
EAST MERSEA
Old Hall Marshes
VIRLEY CHANNEL
TOLLESBURY
St. Peter's on the Wall
BRADWELL
RIVER BLACKWATER
TILLINGHAM
TEEPLE
DENGIE HUNDRED
North Wyke Farm
Holliwell Point
BURNHAM ON CROUCH
RIVER CROUCH
FOULNESS ISLAND
WALLASEA ISLAND
RIVER ROACH

Thames mudflats

Chapter 4
Food & Fowl

I wake in the half-light. It's the most serene view from any bed in the eastern region. I'm in the Old White Harte at Burnham, and the top-floor room has a box window overlooking the river on which a flock of moored yachts with hulls of many colours are facing the tide like geese grazing on a marsh. Wisps of mist form and vanish in the still air. Across the Crouch, the green walls of Wallasea island grow distinct, and so on the horizon do the tower blocks of Southend. The contrast with the previous evening could not be greater, when breathtaking light came from the falling sun and bounced up from the shimmering water. I felt that I had come to another planet, such was the intensity. A little earlier, I had left Ken on the far bank after hitching a ride across the river in a small tender. We arrived to find that the advertised ferry was a broken failure. So much for the phone calls and arrangements. We then found a monosyllabic harbourmaster bidding to be the most unhelpful person on the whole coast. Once there were bells on posts at both Creeksea and Wallasea, and a few rings would bring the ferryman, even in the depths of night. Not today. But then two boating heroes, Steve and Jeanette of inland Chelmsford, appear at the pontoons just, it seems, as I need to ask someone for a lift. Yes, they say, and bring me across the glittering river with its salty breeze. We stop at the jetty right in front of the hotel.

When I am in the shower, washing away the heat and dust and sweat of the day, I hear strains of a hymn. At first, I assume it's the television, but then crouch in the box seat and peer out of the window onto a ceremony unfolding on the narrow promenade. Beneath is a mix of the mainly elderly and rather young listening to the blessing of a new lifeboat. They join in another hymn: "For Those In Peril On The Sea". Another inshore boat stands off in the river, the lifeboatmen suited and at attention, listening intently. The new orange lifeboat shines, and now the sermon is about the Sea of Galilee and its dangerous waters, also found here at the mouth of the Crouch. And everyone is encouraged to think about those brave men and women who volunteer to rescue others, who often have been careless, sometimes reckless, but who can still rely on the support of coastal communities. Such are the responsibilities assumed by people of the shore. It's part of their identity.

Later I stand outside the pub, leaning on the wall and watching the inky river transformed by the silver moon. The chatter from the bar leaks out through the open windows. "If the evening be fine and warm," said Robert Louis Stevenson in his *Walking Tours*, "there is nothing better in life than to lounge before the inn door in the sunset, or lean over the parapet ... to watch the weeds and quick fishes." Now the barmaid, Amanda, designer and printer by day, strolls out on her break, and tells of a man who recently rowed back to his moored boat after a few too many drinks. His absence was only marked the next day, but no one knew what happened. His body turned up two weeks later over the sea in the Netherlands. It was also from the tap room here that smuggler Fiddler Haynes was arrested 200 years ago, and later executed near Greenhithe. Grisly ends. Amanda says she loves knowing so many people in Burnham, and yet in five minutes can be in the marshes and see no one for hours. She then talks of dashing over to Steeple at the end of the evening to see her parents, and I cannot quite grasp how small the land is to someone in a car. It will take me all day tomorrow to walk there. The next morning, I listen over breakfast to eager sailors talking through imagined races and victories to come. It's Burnham week, and later I'll be chased down the Crouch by towering multicoloured spinnakers.

Before long, I am on the sea wall of the south Dengie and entering what many consider to be quintessential Essex. This peninsula makes up the Dengie Hundred, which retains its Saxon administrative title, with the thin and straight Crouch to the south, and the incomparable Blackwater, as Arnold Bennett called it, winding and twisting to the north. The Dengie is 100 square miles surrounded on three sides by sea walls, massively breached in 1953, the sea water revealing the good sense of farmers' ancestors who located farmhouses on slight rises inland. Today, the Dengie is top-quality farmland, and is home to farmers who are trying to develop new ways of telling stories about the distinctiveness of their produce. Janet Thorogood and Mary Hull of the Women's Farming Union have offered to meet me with a picnic by the sea wall. It's August bank holiday, and they are bringing local food to celebrate this walk and their wild peninsula.

But first I have a dozen miles to walk. I head into the sun, and then watch it swing away to the right. Boats, hull-down, begin to race behind me. I listen to the susurrus of reedbeds on the inside and lap of wavelets on the outside. On the far bank is

Memorial gate, Dengie seawall

Wallasea and its new wetlands, the mouth of the Roach, and the northern Foulness sea wall. Then the glimmering sea proper appears, streaking out to its distant marbled horizon. At Holliwell Point I come upon the brooding multi-storey mine-watching tower, with slit eyes and pointed roof, set half below the sea wall, showing how much the wall has since been raised. Along the whole coast, there's nothing quite so sinister as this abandoned command post that could accommodate twenty servicemen. Looking back I see the vast cornflower sky is stretched with condensation trails arrowing west towards London's airports, and beneath the river wall is a line of emerald green. As the land switches north, the clay sea wall becomes welcome concrete covered with blooms of orange *Xanthoria* lichen, proof of the unpolluted air here by what all fishermen used to call the Main. Says Annie Dillard, "Our lives, our tiny lives. Under these skies."

And then blow me, another dog incident. A middle-aged couple approach.

"What a nice day," says one, sweeping a proprietorial arm across the sea.

"Don't worry about him," says the woman, pointing at the dog. "He's just a bit excitable."

As they pass, it swerves across and nips me on the ankle, and skips back in front of them. I'm tempted to use a walking pole as a spit.

Ahead I come to another surprising shrine. A metal gate across the sea wall is covered with flotsam and jetsam tied to the bars: single shoes, punctured balls, hats, nets and ribbons fade in the sun and salt. Nothing to explain why. I call to an elderly fisherman standing down on the beach and ask about the gate. He looks at me. He doesn't know what it's all about, says he hasn't noticed. I ask about the fishing instead, and he says he is after sea bass; his son is a tiny dot several hundred metres up the beach, fishing too. Sea bass make good eating, but the two fishermen seem to be here as much for the solitude as for the wild food. When Sebald saw fishermen camping on the beach at Pakefield in Suffolk, each alone and equidistant from one another, he thought, "They just want to be in a place where they have the world behind them, and before them nothing but emptiness." But fishermen don't see the sea as empty. They look with hope. This radiant coast is a source of food from the sea as well as the land. Harvey Benham writes there were working fish traps out here until the 1890s. These kettles were typically 150 yards long and wide, with stakes 7 feet high, channelling plaice, dabs, flounder and sole to a 10 foot pocket at one corner. The old boy reminds me to look for the yellow horned-poppies, and I agree to search, and press on, cereal fields to my left, saltings and beaches of shingle and shells to the right. Near by are patches of waxy sea holly, the roots of which were once dug up to make candied eryngo.

On maps, there is almost nothing to distinguish mile upon mile of sea wall, farmland and drainage ditches. They make this out to be an empty quarter. On the ground, it is quite different. I meet a young father and his daughter in pink coat, their bicycles propped by the wall. He stops his conversation on a mobile, and we chat about the coast and my walk, days and miles to come.

"North Norfolk?," he queries, shaking his head. I wonder about their day, out here, miles from what some call civilisation. "Just an escape," he says. "Lovely day."

Inland is a mosaic of colours, stubble after harvest, wheat still to be combined, lucerne, small copses, drainage ditches, hay meadows, distant farmhouses and barns. I know there are rare bumble bees, the shrill carder that feeds on clovers and horehound, as Ted Benton has recently written in his definitive work on bees in Essex. There is little in the vertical. No pylons or towering trees. The regular outfalls show this is both a drained and named land – Coate, Ridgewick, Grange, Howe and Glebe. But now I am running late, and press on. Just when I think I am near the picnic place is a diversion inland along a straight dyke and back out again towards saltings on the seaward side.

"We were worried," says Kate Howie.

She and Jo Partridge have come out on a search party, and we walk together to the tartan blankets laid with hampers in the lee of rippling grasses on the sea wall. Here are Janet and Pip Thorogood, Mary Hull, Jane and Michael Pudney from the St Lawrence uplands, livestock farmer Jo, Sam and Rose Lonigan and son Giles, Sally and Steven Green, organic-vegetable growers, Jill and Roger Scott, and Anne Symes, whose farm is inland from where we sit. And it is heaven. I could stop all day, given half a chance. Eat and sleep, perhaps. Listen to the stories. Lunch is wheat rolls made from local feed wheat, chicken, local apples and tomatoes, blackberry-and-plum crumble, flapjacks, goat's milk ice-cream, pear and apple juice. A great celebration of the local. On these Dengie and Tillingham farms, similarly great collective meals used to be cooked up. Each farm worker brought his own contribution to the farmer's wife, who would make up a gigantic stew of hares, rabbits, oxbirds (dunlin), waders, eels, swedes, turnips and fish. The men, popularly thought to be on the run from the inland police, flocked to the remote marsh farms of the Essex archipelago when labour was in demand. Another favourite marsh meal was oxbird pudden, flavoured with pork fat, and also containing moorhen, pewits and tukies (redshank). "A work of art of infinite flavour," marvelled James Wentworth Day. But I suspect these feasts were uncommon. Samuel Bensusan, author of 500 stories from the Dengie, said the conditions for most farm workers were "not far removed from slavery", and any form of meat was a luxury, except perhaps at harvest time.

I ask what everyone thinks is special about the Dengie. There is no hesitation. This granular land is both wild and used. There are further signs of multiple use. I later walk past a star-shaped decoy pond and the distant chapel at Tillingham belonging to the Peculiar People. To the seaward side is a row of barges recently sunk to help capture silt and mud and the remains of fish kettles. In 1594, John Norden wrote that "Essex is mostly fatt, fruteful and full of profitable thinges, exceeding anie other shire." In these same liminal landscapes where land and water daily intersect, the marshes and saltings bring a distinctive flavour to a land of food. On an autumn day, there would be the sounds of burbling curlew, piping redshank, clamouring geese and whistling wigeon, all birds once regularly eaten. In the creeks are oysters and migratory eels, swimming free as tastes have changed. On these walls are sea beet and purslane, crisp and tangy in salads, and out on the saltings

St Peter's Chapel

glasswort, often called samphire, now resurgent in fish restaurants and "redolent of iodine and sea breezes", as Richard Mabey has written. On reclaimed grasslands are wiry marsh sheep, and behind the borrow dykes, modern cereal fields. These remote places are defined in part by their unshakeable wildness, but also by the foods we obtain from them.

I arrive at Janet and Pip Thorogood's farm at North Wyke having walked these same sea walls and saltings since before a winter's dawn. We sit in the cosy farm kitchen with views over a wide land dipping down to the sea, and share a farmhouse breakfast. Being cut off from the rest of Essex, from the rest of England, is the essence of Dengie. It is a land of narrow roads and sharp right-angled bends, a place illuminated on three sides by the light from the sea as well as the sky. There was a not so distant time when every local family was tied closely to the land. When Janet and Pip started farming here in the 1950s, there were seventeen men permanently employed on their 600-acre farm, and thirty casual women taken on at high seasons. There were still horses, shires and Suffolks, and most farms were mixed enterprises of dairy cattle, cereals, vegetables and sheep. Stackie barges took the hay and other products to London, and came back with wastes to put on the land. Twenty years before, rural Essex had been derelict, and landowners were desperate to persuade farmers to immigrate and make the land productive. Some came from Lancashire and the Lake District, and others from Scotland. Landlords offered peppercorn rents if tenants promised to farm properly and put the land in good heart. Janet's mother's family packed their whole Ayr farm on a train, livestock, machinery, furniture and all, and moved here. But there was still not enough local work, and some agricultural labouring families went north. Dengie Hundred coal miners originating from Tillingham, Southminster, Bradwell and Mayland came to be well known in Durham. Today, just Pip and his brother work this farm. It's no wonder that farming has become more distant from its consumers, who tend anyway to think that food comes from the supermarket rather than the land.

There are few owner-occupiers on the Dengie. Most land is owned by St Paul's in London, the cathedral for all of London and Essex until the nineteenth century, and by some colleges at Oxford. It's been in their hands for approaching 1,000 years. At the high point of the farm, you can see down to the Thames and across to the North Downs in Kent. Nearer, the Crouch lie 3 miles to the south, and the Blackwater 3 miles to the north. Straight ahead, it's 4 miles to the North Sea. Today, it's tranquil and quiet, with big skies that reach down and open your soul. They say you don't feel hemmed in. Once though, there were 50-foot elm trees in all the hedgerows before Dutch elm disease struck. With them went some folklore. "The mos' unfriendly tree in the world is an ellum," said one local resident to Bensusan. Never stop under one in a storm, for "they will throw out a branch out of mischief". The Dengie is also the location for J. A. Baker's book of the dead, in which, as Rob Macfarlane writes, so little happens over and over again, and yet the sabring stoops of the peregrine rip and tear and leave fear across the whole of the land.

Wild birds have always been food for people along the coast. Farmers hunted the commons, as did farm workers, fishermen and craftsmen. One of the most famous wildfowlers lived up ahead of me on the north Dengie. From several miles distant, you can see the red tiles and grey ragstone of the simple chapel at St Peter's on the Wall, and around to the west the rectangular glass- and concrete-blocks of Bradwell nuclear power station. St Peter's dates from just 200 years after the Romans left, when Cedd came to re-establish Christianity on these shores. Why he chose such a remote place is not clear. Perhaps it was the wildness and lack of local opposition, or maybe the plentiful brick and stone available from the abandoned Roman shore fort of Othona. People still come from far to walk down to the chapel and its austere interior, for special evenings of song, and for the annual pilgrimage. They come too to walk across the saltings nature reserve at Bradwell Cockle Point to the rare strip of bright white beach between land and sky.

Most visitors miss a small cottage of creosoted barge-boards and ships' timbers hidden in a windswept orchard of apples and blackthorn just over the wall. This three-roomed cottage was built in the 1750s to house the coastguard and his family, and later tenanted by the Linnets. Walter, described by Wentworth Day as "shy and silent, a strong man without fear", was the last to live here, and it is now used by the Wildlife Trust and local birdwatchers. The Linnets were the last professional wildfowling family on the Dengie and Blackwater. Walter worked into his 80s in the late 1950s, his life dominated by the tides and birds. He once led a mass shoot of Brent geese, in which 31 of 32 punt gunners discharged their guns together and brought down 704 geese. On another occasion he shot a flamingo on these marshes. Said Charlie Stamp, professional wildfowler from Canvey Point, and believer in ghosts and wizards, the bird was "a yard an' a harf high. Linnet was arter him for a week before he copt it." Another Blackwater man, Gunner Cook, spent his life sailing a small smack around the region's creeks, switching to the punt for the hunting. These punt guns had 7- or 8-eight foot barrels, and fired three-quarters of a pound of shot. The barrel rested on the bow, the stock on the gun beam, and punters had to creep up close to the fowl before firing. Everything shot was eaten or sold, often to markets in London.

But stalking could never match the region's decoy ponds for numbers of birds. These were introduced by the Dutch: the word decoy itself comes from the Dutch for duck cage, *eenden kooi*, abbreviated over time to decoy, and later taking the meaning "to lure". For a couple of hundred years, duck decoys were a vibrant part of the rural economy of the coast, luring in duck that would end up in butchers' shops. At their height, there were twenty-nine in Essex, fourteen in Suffolk and twenty-six in Norfolk. Decoys were dug with a central area of open water and curved channels called pipes some 50 or 60 sixty metres long radiating towards tapered ends. The pipes were covered with netting made from sisal or hemp, never wire as it whistles in the wind. The trick was to make use of the ducks' natural response to shore predators such as foxes: they swim towards them to investigate, and then stop at a safe distance.

Decoy dogs were usually named Piper. Small and reddish in colour with a bushy tail, rather like a fox, they were trained to attract ducks into the pipes. Tame call ducks were also put out on the water to attract the wild ones, and had notches cut into their webbing so that they could be identified at the end of the pipe. October to March was the season for decoy activity, and some caught 5,000 birds a year, mainly mallard, teal and wigeon. Along the coast, there were decoys at Tillingham, Goldhanger, Old Hall, Pennyhole Fleet, Nacton, Benacre, Iken, Flixton, Fritton, and Dersingham. In the twentieth century, the use of decoys declined, as wetlands were drained, and as domesticated meat became common and cheaper. The last commercial decoy was at Nacton on the Orwell, lasting from 1830 to 1968. It is now run for ringing purposes, and reveals what extraordinary journeys these migratory birds make: pintails have come in from Russia, teal from Scandinavia, wigeon from Siberia, and mallard from Canada.

- - - - - - - - - - - - - - - -

One sun-washed autumn afternoon, I head back to Bensusan's peninsula, driving south into Dengie's narrowing roads and then on to the very edge. To the left the Blackwater is a deep blue strip beyond the fields. I meet David Reed in Bradwell, and we load up a Land-Rover reeking of marsh, and are off to the upper Crouch, the black Labrador in the back tense with anticipation. At the marina we pull on mud-stained waders, jackets, hats with face net, and then we are in the dinghy, outboard motor whirring, and briskly moving east downriver and then south inside the island's old river wall after Easter Reach and Althorne Creek. We're now inside Bridgemarsh Island, the river wall a thin broken strip of eroding Kentish ragstone topped with mud and grass, long since breached. It was abandoned in the 1930s. The final occupants, the Gooches, lived in the upper rooms of their lonely farmhouse as the tides washed through the ground floor twice a day. Wentworth Day knew this drowned island in the 1950s, when it was alive with sheep and cattle, the dykes and fleets home to vast numbers of eel and duck. At the west end was still the old farmhouse, by that time ruined and empty. But then the sea wall was breached again and all the livestock drowned. The sun is strong on this clear October afternoon, and I feel the warmth on my cheeks and hands. Later, it will become piercingly cold.

At a tiny hide we put out the netting, and settle down to watch the light change, the tide rise, the birds come and go, and the marshes gradually dim from dazzling daylight to bruised dusk. Eventually the dark rises and takes over the world. It's so very close to the rest of civilisation, yet we are apart. A train runs soundlessly along the south Dengie shore. Silent traffic on the hilltop weaves among houses and churches on the skyline. Out here we can hear the plaintive curlew, liquid waders, "teu-eus" of redshank, the whispering laughs of wigeon and all the while the water laps and gurgles as the fleets, creeks and rills fill with the returning tide. A marsh harrier quarters the saltings, a dark and deadly shadow, bringing a predator's temporary silence.

Then wigeon are glittering towards us in the cloudless sky, in fast twisting flight, the whistle of the drakes like a "phew" of surprise, as Julian Tennyson (great grandson of the poet) called it, their white flashes accompanied by arrows of black wings, whiffling in as they see the decoy ducks in the creek. And, bang, bang, Dave has leapt up after we've sat motionless for so long. Then a dark fluttering, a gliding in, and a duck thumps into the glasswort and cordgrass, and off goes the glossy dog, leaping and quartering the land, returning with duck held carefully in mouth. He drops the wigeon and shivers off the marsh water, sickly-sweet, the spray of droplets catching the sunlight. Two minutes of intense activity are followed by more stillness and waiting. Dave talks of his former life as a lighterman on the Thames, like his father, then stretches as a city policeman in Poplar and Tottenham, ending his career with the rank of Chief Inspector. And now, strangely, out here on the marshes, our lifeways connect. He and fellow coppers used to drink at the Still and Star in Aldgate, my mother's grandfather's pub. The Blood Inn, they used to call it, he says, but didn't know why. I tell him there had been an abattoir next door. It made my grandmother a vegetarian. He tells of investigating dreadfully painful murders of children, a suicide on iron railings outside a house, car crashes, violent crimes, how these events snap and fix in the mind, and how impossible it grew to pass those places without the memories rearing up. The city became a place of sorrow as the years passed. But out here by the glassy water, wide marshes, and vast sky, new memories are only of immersion in the life of ducks and other birds. Dave cannot imagine doing anything else now.

The gleaming light fades slowly to pink, and the sun boils into the hill behind us, where the dark silhouette of Canewdon Church marks the nearby site where Cnut defeated Edmund Ironside in 1016, and was declared King of all England. Where did he, known today as Canute, stand and try to hold back the sea? Perhaps here at

Bridgemarsh Island

Bridgemarsh? The light falls further, turning the saltings from pink to yellow, then to darkening green. The half-moon brightens, and con trails streak across the amethyst sky. The other world goes onward the same, as Hardy wrote, but this marshland dynasty will not pass. No one would know we are down here in such acute beauty. The cold now rolls over the marshes and across the carpets of seablite, glasswort and sea lavender. The huge skies stretch from one era to another.

The light dies, and we have two wigeon. And we've been left alone by Canewdon's witches. We climb into the boat, pull in the decoys, the dog satisfied and curled up against my legs. An icy cold envelops us. I put my gloved hands inside the tops of the waders, and hunch forward. We glide the couple of miles back, followed at first by a yacht with a single light on the mast, but then we are on our own in the black creek and heading towards the sodium lamps of the marina. This marshy afternoon has been dominated by the intensity of watching for birds, the listening, the caresses of sun and cold, the sweet mud, and lonely saltings. All bleak and beautiful, remote and yet near by, a soundscape of piping, bubbling, crying birdsong. Later, at home, the wigeon are laid out on a newspaper, the male with golden crest. Then a carpet of white surface-feathers and soft inner down. The odour of the salt marsh is upon them, and clings to the kitchen. Part of a long tradition of food from the marsh, life from the land and the sea. Wigeon can be fishy if they have fed off eelgrass, but these have the flavour of inland marshes.

I turned the corner at St Peter's, passing through the cool groves of the Othona Community, but without stopping that day. I return to meet manager David Williams and colleagues Matthew Dell and Gail Hayne on a still autumn day with a sweet smell of damp soil and decaying leaves in the air. Remoteness from centres of power and cultural norms often brings Nonconformism, and religious difference is one of the features of this coast. One small but notable movement was the Peculiar People of the mid-nineteenth century. It began in Rochford, and eventually spread to forty-three chapels across south Essex and east London. James Banyard was the founder, and had attended Wesleyan chapel until the late 1830s, but then broke away from the main body of Methodists by preaching on Mondays and at other irregular times of the week. He and his followers faced immediate hostility, rubbish and stones being thrown at their homes. They believed in the absolute healing power of prayer, and this invited further controversy. The first incident occurred in 1848 at Prittlewell when a doctor was not called and a child died, resulting in denouncing of the Peculiars. Chapels were eventually to be built across the region from Fobbing to Foulness, and Grays to Stanway.

Schisms were to dog them until their dissolution in 1956 when they joined the Fellowship of Independent Evangelical Churches. 1853 saw the first split, after one of Banyard's own sons was taken ill, and he promptly called a doctor who effected a cure. Banyard was to die in 1863 estranged from the rest of the Peculiars. Through the 1880s to early 1900s many were fined or imprisoned for not calling doctors to ill children. By the 1920s chapels had begun to close. Though the movement lasted just a century, it left a unique cultural mark on the county.

A quite different form of Christianity, however, is here at the Othona Community on the north Dengie coast. It has an inclusive rather than exclusive ethic, and set out from the 1940s to be a meeting place for people of all faiths and none. Othona is hidden in a small wood a few hundred metres from the chapel of St Peter's, but half a century ago this was open fields. After the war former RAF chaplain, Norman Motley, had run Answer Back sessions for servicemen in London, and then the Nails

Movement to encourage debate and discussion on international relations. He travelled widely in search of a place to establish a retreat, and having visited Iona, Lindisfarne and Glastonbury, was posted to Stisted parish in Essex. He was persuaded to visit St Peter's in 1946, just as all the mines had been removed from the area. The chapel immediately captivated him with its overpowering "sense of thirteen centuries of prayer", and the decision was made.

The Army said yes, and Motley was given permission to use the Nissen huts near the sea wall. In the first summer German prisoners of war, Russians and Armenians, refugees, students, parishioners from Stisted, and people from London and south Essex came to join him. From the start, he believed that Othona should engage in renewal and reconciliation, and the place itself played a vital role. People shared in the simple pleasures of walking, swimming, routine tasks and relaxing. Motley felt that "the great skyscape and the spread of the saltings [brought] an immense degree of peace to mind and heart". In 1961, Othona bought neighbouring East Hall Farm, thus securing the community's future. In 1992 some of the old buildings were demolished, and two years later the new ones completed. Waste water is now processed in reedbeds, and food sourced locally. It has retained its magic. One visitor, Ann Froggett, has written, "It was like nowhere I'd ever been, even knew of, ever imagined," and another, Jean Pullen, said, "I like leaving the everyday world behind for a while, the simple lifestyle ..., the fresh air and sea wall walks, having interesting company, having time to listen and talk."

Over lunch, we talk though Motley's vision. In all, there are 600 members of Othona, and about 100 come to stay at various times of the year. Some come for a working weekend, others feel Othona is their second home. Some first-time visitors are unnerved by the open skies, but most say the tranquillity helps them recharge their batteries, a metaphor that suggests modern life drains them. In one sense, it is a place to which people can escape, leaving behind their other lives. It is also a place where new norms can be created.

It is because of the uniqueness of both Othona and St Peter's Chapel that the annual Bradwell pilgrimage attracts 1,000 or more people to this coast in early July. The following summer, I sit on the bench by Linnet's Cottage and look out across the saltings sprinkled with the mauve of sea lavender and crimson of sea blite. To the south, a heat haze ripples across the sea wall and marsh; to the left and north, 3 miles away, are the distant shores of north Essex. White horses dot the blue estuary mouth. How appealing is another coast when seen like this. Bensusan, the laureate of the Essex marshes, looked across here, and wrote that the "mysterious country across the estuary enshrines more possibilities than Tibet". At that time, there were local "villagers who have passed more than eighty years on this side of the Blackwater [and] have never been across the estuary". Speaking for himself, he wisely reflected that, "The one fear is disenchantment. Suppose my expectations should not be realised?" Perhaps these views always work best from the southern shores looking north, with the sun over your shoulder. The sky ahead is darker, and the marble white of sails and the multicoloured jumble of houses pick up the sun.

I walk from the coast back up to Bradwell village, past the desolate house-base where a man recently took a bulldozer to his own house in a domestic dispute. St Thomas's Church is packed full, as is the churchyard. I meet Reverend Paul Trathen and a group of his Rawreth parishioners. He's wearing a black cassock, bush hat and sparkling earring, and describes how he's persuaded a supermarket chain to put their hands in their pockets for major repairs. I meet the Bishop, who's meeting, greeting and moving swiftly on as the band plays. He says a few words in the churchyard, welcoming pilgrims from very far and near. Bell-ringers gaze out of the tower windows, and then in a great rush everyone is off.

Paul and I talk as we walk to the chapel. I point out the burning bushes. All the elms that have resprouted after previous attacks of the Dutch elm fungus have been hit again this month as if by flame thrower. Every hedgerow on the Dengie is picked out with patches of crisp, browned leaves. The previous elms still remain, dead and bare. But this generation is going too. The very last tree before the chapel is, I know, a great elm. It's one of the few mature specimens to have escaped previous attacks, and it's a multi-trunked survivor. But when we look closely, we can see the leaves on one branch are dried and bronzed. A flock of goldfinches chatter by, swifts zip across the fields, and skylarks hover and sing to the blue sky and racing cumulus clouds. The day had started grey and quiet, but by mid-afternoon the wind is buffeting, flipping over leaves that flash light green, and the crowd is sitting and milling, watching and waiting. There are two large marquees by the 55-foot chapel. The canvas cracks and slaps in the wind, the wooden posts creak, and the heat builds. The grass is fragrant. I speak on our growing disconnections from nature and place. People nod, and at the end wander off for the outdoor ceremony against the ancient chapel that has variously been beacon tower, smugglers' lair, orderly room for troops, cattle barn and stable. I slip away through the woods.

- - - - - - - - - - - - - - - -

On the long walk, I had left Othona, and walked to and beyond the twin blocks of the decommissioned power station. The sun is dropping into the upper reaches of the estuary when silver-haired Michael Pudney picks me up from Bradwell Marina. The sea wall was hot and dusty, and I now gently subside into a comfortable car seat after nine hours and 18 miles. How miraculous to be able to move in a car from one place to another with so little effort. Across the azure water lies the island of Mersea, fringed with coloured beach huts, and far to the right the Tendring shore. I know it will take me two more days of walking to reach Mersea, as I will next turn west and head inland along one of the saltiest estuaries in the country. Michael runs us up to St Lawrence Hill in the middle of the Dengie to see his best view in Essex, the sweeping Blackwater ahead of us. Yachts flash down the estuary, heeled over, zipping across through the waves. The salt-making centre of Maldon is on its hill to the far west, Tollesbury Marsh on the far side, Steeple Bay in front, where the Krays stayed in their mother's caravan, and Bradwell over to the east. But a car has nothing on a plane, for before dropping me at the Star in Steeple for my overnight stay, Michael invites me to join him on an evening flight. What good fortune is this. After running methodically through the flight checks, he guides the twin-seater past teetering hay bales on his farm's airstrip, and we rumble forward and lurch into the air.

The fifty-minute flight takes us up over the Blackwater, past Osea and Northey Islands, where the tide is lower than Michael has ever seen it. We fly back down the estuary, and then around the aerial exclusion zone over the power station, and past spits and mudflats far off to sea that are making a rare appearance. From here we can see that the land is all sinuous, meandering creeks, twisting rivers, curving borrow dykes, and saltings cut with dendritic rills that seem to reach out for synaptic connections. We see the spectacular effect of the ten barges sunk in a line to the east of our lunchtime picnic site, a square of new salting appearing where the silt has settled. We run south towards Foulness, over Fishermen's and Asplin's Heads and then to Wakering Stairs, and Michael points to where his grandfather used to farm on Foulness as a tenant. He made a small fortune selling hay to the Army in the First War. Then we dip a wing westwards towards the sun setting into the Thames. And the colours explode. Incandescent reds and burning golds light up miles of mudflats and shallow surface water off the shore. Southend and its silhouetted pier can rarely have looked better. Far in the distance, we see the towers at Canary Wharf in the fiery smog. All the land and sea from here to London is ablaze. I wish we could hover here forever, but within a few minutes, the plane leaves these glorious muds, and swings north towards Maldon, the three parallel rivers of the Roach, Crouch and Blackwater ahead, and then back to land on the bumpy grass strip. In minutes, we have passed over a land that I have taken days to tramp. We sit and catch our breath in the garden, glasses of white wine at hand, and talk about that sunset and the successful harvest. Life at the farm looks much better now that cereal prices are up. And the weather came good in the end too.

For two days I continue up the Blackwater and down the far side with Janie Wilson joining me on the first day, and Jo Barton and Sarah Pilgrim on the second. Hazlitt and Stevenson would have disapproved. They preferred to walk alone, but I'll enjoy the company. Janie's an old college friend, travels the world, and drives up from south London that morning. We park at Sally Kirk's goat farm at Mundon Hall. She has 1,000 sleek white animals housed in a barn covering an acre, together with a small herd of alpaca roaming incongruously across the fields. The dairy is sparkling, and the ice-cream smooth and creamy. Sally, though, still wonders how they can make a proper living. It is easy to say connect more closely with consumers, but how to do it and make enough money to survive on these tenanted lands? We have no answers, but we're here, too, to see Mundon's uniquely petrified forest of ancient oaks.

In a single patch of her parkland are twenty-five trees beseeching unseen gods with long tapering branches. Their trunks are sculptured like driftwood and most branches have no leaves. No one knows quite how long they have been here, perhaps 1,000 years? But if it has been so long, it is not clear why someone did not cut them down for their valuable wood. Who felt they were important enough to keep? What magic protected them? We find that two of the trees retain some fissured bark and are sprouting green side branches, and these are heavy with acorns in this already notable mast year. Half a century ago, Wentworth Day described them as "great stag-horned oaks in an old rough pasture like a little park", which implies they were living trees then. Sally says they look best in the rain, when the woodgrain is accentuated, and the trees seem more alive. Not many pass this way, and the trees remain something of a secret, though inveterate tree-climber, Rob Macfarlane, came here too in his recent journeys through Britain's *Wild Places*. But these remarkable trees are no obvious help to Sally and Bob in their farming, and it's

Mundon oaks

surprising that this field has never been turned into farmed land.

The south shore sea walls are dense with tall grass. Not many have been this way, and the walking is hard. We talk of home, kids, growing up; how have we become so old? In the Blackwater is Osea Island, with a few houses and a dozen permanently dry fields, and Northey, mostly saltings covered at high tide. It was in these muds and creeks that twenty-four revenue men had their throats cut in the smuggling wars, and no culprits were found. In the late 1940s Mayland wildfowler Stanley Tiffin came upon a rare harvest while paddling his punt: a parcel containing the torso of a man, later identified as Stanley Shetty, car dealer from the East End. It was a national sensation, and eventually a member of Elsdon Flying Club was arrested, tried and found guilty. He had chosen to drop the body into the marshes rather than bother to fly further out to sea. We pass the site of more gory history, where ninety-three ships led by a certain Olaf battled their way up from Kent in AD 991, and fought the Saxons under Byrhtnoth at Maldon. The Saxons foolishly allowed the Danes to form up on Northey Island, and march across the narrow wetroad before engaging them in battle. Byrhtnoth was killed and the Saxons routed.

Heybridge shore

We're sneaking into Maldon and all of a sudden come to the recreation ground, and amongst the pools and fountains are hundreds of people. So many aggregated together, as if for safety on these soils soaked with Saxon blood, and yet near by we were in places so completely deserted. Canoeists stand by their orange- and red-vessels, holding bright-blue paddles, and beyond them are the wooden Thames barges at Hythe Quay, brick-red sails tied up, and the busy boatyards resounding with hammers and drills. Then up into the town and bustling main street, past the triangular tower of All Saints with its George Washington window presented by the state of Massachusetts, and down into Heybridge district of converted mills and roundabouts drowned in traffic. Maldon is famous for its salt-making, the process heavily promoted by the Roman economic machine. There were once forty-five salt pans around the town, and all along this estuary are many Red Hills, made up of the briquetage of broken ceramic and terracotta pots used in salt-making, some dating back long before the Romans. Good people are still, of course, worth their salt.

The canal navigation is quiet. We watch flitting dragonflies above the limpid waters, and then come to another honeypot at Heybridge Basin's lock-end, with its clapboard pubs and houses. There are plans for development here, and also notices of opposition. Save our village; say no to Lock Hill development. We walk on past a chandlery with flaking paint, wooden huts with asbestos roofs converted to tiny homes, ranks of moored yachts, and two grandmothers sitting on a concrete wall spearing chips, two teenage girls beside them talking into their mobiles. The older two smile, but the younger pair are in another world. Inside the sea wall caravan parks sprawl, permanent new towns out here on the marshes. Then the Osea wetroad magically appears as the tide recedes, and the island is temporarily bathed in bright sunlight as fearsome black clouds bank up and arrow towards us.

Osea was purchased in Victorian times for its curative bathing and bracing sea air, the plan being to establish a colony for city alcoholics. It didn't happen, though a motor-torpedo-boat centre was set up there during the war. We pass another former decoy on Gore Creek, then turn inland up a green lane to find the Chequers pub, a former smugglers' haunt in Goldhanger, just as the clouds lift again. We wedge into seats in the empty bar, wondering for a moment whether they are reserved by particular locals, and enjoy the pleasure of a cold beer. Janie heads back for Mundon and London, off for a life in France before long, and I settle in for the night as the pub's sole guest.

- - - - - - - - - - - - - - - -

The breakfast is far too good. All from local produce and temptation wins. Jo and Sarah arrive, and we three set off for a final day on the Blackwater. We'll be walking out to Tollesbury and aiming to finish at East Mersea. The Blackwater's magic continues in the bright sunlight from sky and water. Piping waders echo across the saltings, some covered with stands of golden samphire. A fisherman brings his blue boat across the muds by tractor perched up on four super-inflated tyres. We walk on, at times directly towards Bradwell, at others back inland up creeks. We pass one where half a century ago a man drowned in the mud. Wentworth-Day records wildfowler Alf Claydon's tragic testimony: "He hollered all right – for an hour or more till the tide covered him. When they found him in the morning, he'd scraped a great hole in the mud big enough to hold a donkey cart. An' he was up to his shoulders in mud in the middle of it. Drowned! Pore chap!" After miles of green sea walls and half a dozen more Red Hills, we switch inland towards Tollesbury, and notice how the air temperature leaps when the water is left behind.

The hedges are heavy with red berries of hawthorn and hips and the blue-black

Mersea beach huts

of sloes. Such abundance used to presage a bad winter, although such folklore weather predictions are less likely to work in these times of change. We will, as it happens, have only two brief snowfalls this coming winter, and those are going to be in late March and early April. In the old days, self-contained Tollesbury had blacksmiths, bricklayers, shoemakers, thatchers, a watchmaker, wheelwrights, seawallers, saddlers, millers, grocers, bakers, a butcher, coal merchant, fruit dealer, wildfowlers, and hundreds of oystermen and fishermen. Up into the village we walk, then down past a field of snickering horses, and suddenly the land opens up into another of the county's spectacular views. The horizon is rimmed with low hills, the slight rises on Mersea and towards Peldon and Colchester, with a small gap of the sea to the east, and then further to our right the masts of Tollesbury Marina and its red lightship. There is a sense of being in a bowl, with mile upon mile of wet saltings and mud before us. Ahead is Tollesbury Fleet, Old Hall Marshes, and Little and Great Cob Islands. A fringe of dead trees lines a flooded field, and inland a green tractor inverts a stubble field brown as a swarm of gulls jostle. We three stand, as if at the end of a play, silently listening to the piping calls and aching emptiness.

We had intended to arrive on Mersea by oyster boat, but an accident that morning had seen Richard Haward off to hospital. I call Andy May at the Essex Wildlife Trust, and he and Alan Shearring come to Old Hall Marshes to run us around by car. It's another cheat, but a welcome one. The three of us first eat lunch on the river wall as we wait, right in the middle of more smugglers' territory. We are by the strip of white cottages that once was the Hoy Inn, or Ship Ahoy, or just the Ship, as it was variously known. By this old quay, there were also lime kilns, an iron foundry and brickworks. These marshes were first reclaimed by the construction of sea walls in 1598, and took sheep brought by boat for fattening before their final trip to London's wharves. The local decoys were constructed in the late 1600s, and one continued in commercial use until 1914. Now it is all a fine bird reserve. We feel like scratching around a bit. There's a persistent legend of treasure buried somewhere near the old Hoy Inn.

For the rest of the afternoon, we walk the southern shore of Mersea Island past lines of beach huts, ranks of caravans and clumps of sea holly by an astonishingly aquamarine sea. And over there, on the other side of the estuary mouth, distant Bradwell and St Peter's. We're now on the shore Bensusan viewed from Bradwell. A squall passes mid-water, and turns the sea dark. Kids lark in the water, and men on jet skis roar around in circles. We come upon dead animals washed up by the tide. First dozens of jellyfish, then a partial rabbit's skeleton and many pieces of driftwood. Then just as I am recalling that huge rat at Tilbury, we come upon a half metre monster with vicious teeth, folded fore paws, and abdomen swarming with maggots. The drawn skin on its skull and the teeth stained pink by the sea suggest an angry animal. It's a mink, late invader of this region, and voracious predator of water voles. Just past a pillbox slumped in the waves, we climb the low cliffs, and turn inland across the meadows at Cudmore Grove. A grandfather with his younger family watches, turns slowly, and shouts, "You forgot your snow."

N
W
E
S
ESSEX
NOR
FAMBR

Chapter 5

WEST MERSEA
GREAT COB ISLAND
Shinglehead point
Tollesbury Wick Marshes
GOLDHANGER
Bradwell power station
RIVER BLACKWATER
CLIFF REACH
CREEKSEA
LION CREEK
RIVER CROUCH
WALLASEA ISLAND
PAGLESHAM
FOULNESS ISLAND
POTTON ISLAND
THE MIDDLEWAY
OCHFORD

Osea Island wetroad

Chapter 5
Wild Archipelago

The ground at dawn is sprinkled with white frost. It seems autumn has raced into winter. The land is wreathed in mist. "Fog everywhere. Fog up the river… fog down the river… fog on the Essex marshes," wrote Charles Dickens in *Bleak House*. Later, when the smoky banks burn off, and the sun is high, we will wonder how many days in the proper summer were as warm. I arrive in the half-light at North Fambridge on the south Dengie, a ghost settlement by a river that could be a tiny stream or a wide reach. A moored boat grows distinct out by the clubhouse on stilts. A seagull perched on a post in the water remains unmoving as a group of men park their car, form up and disappear down the gangplank. For a while, I have no idea at all whether this is the right place. I walk east along the river wall, the grass still crusty, and the grazing marsh of Blue House Farm emerges. Most of the ditches on the reserve have become populated with rare water voles, and luckily mink have only been seen passing through. I know the sun should come, as the stars were bright from up on the nearby hill. But down here by the river, it feels like we might be isolated forever.

These shrouded marshlands of Essex were the frontline for centuries against the marauding Danes or Vikings, as they were variously known. There were raids, battles, defeats and great victories. Some stayed, one ruled, yet history remembers the Normans of 1066 more. Nearly 1,000 years after Cnut, the sun appears this morning first as a yellow disturbance in the mist, grows and coalesces, and then reflecting on the water becomes a pair. The light now rapidly increases in radiance, suffusing and then repelling the fog banks. I see a powder-blue sky begin to appear and down here the grass turns glistening green and yellow, the mud brown, and the water a pale eggshell blue. Still mist clings on awaiting its dispatch. Ken Worpole writes that this Essex landscape "has to be experienced as terrain, ground to be walked, mud-flats and creeks to be gang-planked or forded, undergrowth to be circumnavigated, rain, wind and sun to be accommodated". Today it'll be by boat, and naturalist Brian Dawson appears wearing his habitual lumberjack shirt, ready to lead this trip. We hear a low throbbing, and an old Series 7 offshore orange-and-blue lifeboat slides up to the pontoon. Tall and wiry, John Rogers lives on this boat and goes where he pleases out in the archipelago. Our small group clambers aboard.

The Crouch runs almost straight from deepest Essex to meet the North Sea between Foulness and the Dengie. As we head east the sun works on the last wispy wraiths clinging to the water. The surface is millpond smooth, glowing, the wake of the boat advancing towards both shores. We motor past Longpole and Shortpole Reaches, then Bridgemarsh Island where I had crouched in a cold hide waiting for flickering wigeon. I can see a pile of bricks, reminder of the industry on this abandoned isle, but there is no sign of Gooches' farmhouse. The outer sea wall is sparse where it failed and let the river in. During spring tides this island is completely covered, and John points to where a boat was stranded on the saltings for weeks after its owner thought he was in deep water. We pass Cliff Reach, a fossil site, and Creeksea with its ferry staithe, and on the south bank is Wallasea Marina where a 2,000 tonne ship loaded with timber dwarfs every other vessel on the water. Opposite is Burnham and its glorious waterfront of hotels and modernist sailing club, houses and old inns. Deep on the marshes to the south you can see this beckoning white strip, and think that there lies a place you want to be.

The river is still gleaming as we pass the five breaches of the Wallasea wetlands, water now pouring out on the ebb tide, and then we are turning south into the Roach and left into the Middleway, a curious creek that links the Thames side of Foulness with the Crouch and inner Essex. At a point where water flowing up from Havengore Bridge meets water from the Roach and the Crouch, the mud piles up to prevent boats from passing. We've arrived. This is the watershed called Devil's Reach, and it is here a colony of common seals have chosen to reside. Today just the seven adults are beached, tails high to balance heavy heads, stationary on the slippery mud. Others in the water emerge and gaze wide-eyed. John drops the clanging anchor, and we sit back to watch. Redshanks call "teu-eu", glide and flip-flap across the muds, and there are curlew, lapwings, dunlin and knot too. Earlier in the mist we had heard the burblings of Brents on the seaward side of Foulness. Wentworth Day said of this Essex archipelago that it "lies bleak, almost treeless and shunned of men … it is a country of the wild goose and the curlew, the wigeon in their mewing thousands, and bouncing great hares". In the 1920s, Wentworth Day held the shooting rights on the 1,000 acres of Potton Island to our right, like Foulness now under military

Mine watching tower

ownership. In those days, there were a couple of houses, three trees, 1,000 sheep and brackish fleets attracting every kind of wader, duck and goose. "The strong smell of the muds," he observed, "and of salt waters, the breath of sea marshes, that indefinable something, when you have known it, will dwell forever in your blood."

Fourteen miles from Fambridge and the water is still a mirror under the warm sun. We watch the seals in silence. Then with no warning they skid with a series of splashes into the water, and bob up around the boat. When we come to depart, they disappear underwater, and we swing south towards Paglesham in search of the site of Darwin's *Beagle*, a recently discovered mark in the mud by an old hut near dozens of rotting timbers. Darwin travelled the world for five years in the 1830s in the *Beagle*. It was eventually brought from Sheerness to the Crouch to combat smuuggling and anchored at the entrance to the Roach as a watch vessel, named *WV* 7. Eventually it was taken ashore at Paglesham in 1863 and broken up for its copper a decade later. Darwin was lucky. Twenty-one of the 107 boats built on the same model as the *Beagle* were lost at sea, and they were known as coffin brigs in the Royal Navy. Today, all that remains are some timbers somewhere in those saltings. The old lifeboat turns about, and on our return, the wind rises slightly, and riffles the river. Now it is busier, and we talk about the cultures of boat people. An official in a red rib patrols, watching for transgressors of the 8 knot speed limit in front of the White Harte. Further along, a water skier leaps and pirouettes, and the Romford navy, casual name for the jet ski set from London, is out too in search of their own piece of freedom. We motor past racing yachts that are struggling in the light breeze. I stand with John as he steers from the wheelhouse, and we wave to other crews as we pass.

We had tried once to find the *Beagle*, but failed. The night had been a dead calm, and a harvest moon had hovered in the trees. Ken Worpole and I left the Foulness Dragon and made for Anne Boleyn's home town of Rochford, the ford of the hunting dogs. We walk from the Boleyn pub down the Roach's south shore, the nettled path following a narrow river clogged with tall reeds. The landscape opens up to grey sky and salt. At the boarded-up Stamford flour mill is a patch of deep-inland salt marsh, and we cross by ligger boards to the north shore. The mill's windows are broken and its walls covered with graffiti, signs of the death of a building. Crossing a manicured cricket pitch we take the river wall towards Paglesham, and are accompanied for an hour by the echoing shouts from a football match up at Stambridge. Over to the south the blocks at Southend are still visible, and beyond them the Isle of Grain chimney too. A blue sail drifts up a creek and back, and up another, leading us towards Paglesham, past Barling Ness on the far side and the western corner of Potton Island. We watch a couple of early geese whiffling in unison, splashing down to join curlew on the muds, and pass an eviscerated gull on the path, its liver lying to one side.

At the boatyard at East End, the sun appears and lifts the day. We are just beginning to think we might ask about the *Beagle* when a man in a tight T-shirt carrying an outboard motor stops ahead. He juts out his chest and jaw, and stabs, "The pub's in the village, and there's the fucking footpath." We stop and look at each other, and back at him. He struts away.

At the white barge-boarded Plough and Sail, home of a celebrity TV chef, there are finches in a garden aviary, and in the dappled light it feels like a rainforest garden. We absorb into the shadows and three separate groups of people come over to chat. We are different: everyone else has come by car. We reflect on these differing receptions as we set off again, across dusty fields first and then back to the river wall, and now we're heading up Paglesham Creek that separates this piece of mainland from Wallasea Island. But we've completely missed the *Beagle*. It's a new day. An afternoon of saturated colour, an azure sky streaked with white cirrus all reflected in the blue water, the dry golden false-oat grasses on the walls, ploughed and harrowed brown fields, others still stubbly-yellow after recent harvest. A lizard scuttles through the grass, and we look out for adders. At a lonesome field maple we rest in its shadow, and then press on as the wall takes us first away from Wallasea, and then back past a caravan site on Lion Creek, all bright-green mown grass. An extended family sits on plastic chairs enjoying this Sunday afternoon, Granny firmly in the shade, barbecue smoking lightly, children climbing a tree. Hurrah, we say. A father drifts a frisbee towards a distant child who skips and leaps. We arrive at the Crouch by the boatyard, looking north towards Creeksea and Burnham, and in a hurry to meet the ferry's boat departure time, only to discover that it's been out of commission for six months.

- - - - - - - - - - - - - - - -

Smuggling unites every creek, estuary, remote beach, marshland and river wall of these three eastern counties. This archipelago was deepest smugglers' country until at least the 1930s. The nineteenth century essayist Charles Lamb called smugglers "the only honest thief", and Kipling's 1906 *A Smuggler's Song* spread more romantic myths. It all started when excise duties were introduced in 1643. At the time, there was no income tax, which was only introduced in 1799, so these duties were vital revenue for government. The 1700s and early 1800s saw smuggling grow to a massive scale, mainly of tea, brandy, tobacco, silks, wool and gin. The revenue authorities were in an almost constant state of war, though many people simply saw smuggling as free trading, and just about everyone on the coast was involved. The 1718 Hovering Act was introduced to make any vessel under 15 tonnes loitering within 2 leagues, roughly 6 miles, of the coast liable to seizure. It seems strange now to think of the likes of tea and wool as having major hidden trade routes and smuggling gangs: the tea cartels, wool rings and gin runners.

New terminology emerged: to smuckle was to conceal goods in a hiding hole, owling the illegal export of wool, creeping the rowing back and forth using grapnel hooks to snag underwater barrels anchored to the riverbed. There were clinker-built carvels, colliers, luggers with three or two masts, bawleys, and sloops. There

River Crouch seawall

were famous smugglers too – William Blythe of Paglesham, known as Hard Apple, was in later life respectable grocer, churchwarden and village cricketer. He died in 1870 and is depicted on the village sign today. At the time the *Beagle* had become a watch vessel, these rivers and coastal waters were major travel and trade routes. Movements of goods by road was tortuous, and flat-bottomed barges transferred agricultural goods from farms to London, and returned horse and human wastes as fertilizer for the land. They carried chalk and lime from Kent for the sea walls, coal in all directions and hay to the cities for horse feed. There were 150 sea-wall quays in Essex where the barges could load and unload directly to and from farms. Barge traffic continued into the first couple of decades of the twentieth century. In amongst this constant motorway of movement, the authorities had to separate the smugglers from the legitimate fishermen, oystermen, bargemen and farmers. Many were one and the same.

There were always strange goings-on out in the wide marsh country. There were wise women, wizards, burning men and witches on broomsticks. Legend says Canewdon always had twelve witches, and Hadleigh and Leigh three each. Wallasea and Foulness Islands shared one witch who rode on a wooden hurdle across the Roach. "Blast boy, she scat me," said one farm worker to James Wentworth Day. "There was several witches in them days, but owd Mother Redcap was the headdest one o' the lot." He worked at Devil's House Farm, but this is no longer marked on maps, one of many marsh farms that have completely disappeared. One of Wentworth Day's finest tales was about one of the last true hunter-gatherers of the Essex marshes. Billy-Boy, as he was called, was a fowler and fisherman who lived in a marsh hut in sight of the wooden cottage of Cunning Murrell, the great wizard of Essex. Billy fished for whitebait, flounder and eels in summer, and shot for a living in winter. In his grey punt he ranged from Canvey to Mucking Flats, and up from Southend pier to Deadman's Point on Foulness. "I never knew him do a day's work on the land", said Wentworth Day, and one day Billy suggested they go gun running across on the marshes of Kent. "We'll go gunning for doocks an' then runnin' from them tew ol' farmers an' their tew great dawwgs. Bit o' spoort." But Billy-Boy warned of the greater risk: "There allus was witches about these here parts. If they witched ye proper yew was a gonner. Yew jus' up an' died." Unless you could get help.

Cunning Murrell was born in 1780 and not at all like Gandalf. He was thin and less than 5 foot tall, and a herbalist, healer, seer and animal whisperer. Arthur Morrison describes him as "quick and alert with a blue-frock coat … and a hard glazed hat. Over his shoulder he carried a large gingham umbrella with whalebone ribs." He slept by day and walked at night, always with a basket for herbs hanging from his umbrella's handle. He lived in Hadleigh, and received a dozen letters a week from people of Suffolk, Essex and north Kent, many about bewitching and losses of clothes, crops and lovers. He asked those who came to him if their problem was high or low: did they require magical or material help? It's easy to imagine night-time on the marshes in those times: no lighting from nearby settlements, just the moon and stars, whistling wind, scudding clouds, animals tame and wild, ghosts and witches imagined in every shadow, and everyone fearing to go near the ruined and haunted castle.

They're like tickertape in the early sunshine. Four hundred lapwing swoop, swirl and twist. Against the blue, they turn and for a second are absorbed. In a blink the sky ripples, and they are dark again and flying as one towards me.

There's a soft whirr of wings as they pass overhead. Something has put them up. Out on the marsh come alarm calls of gulls, harsh, mixed with burbling waders. To my left, out on this river wall, is a gap punched through in 1995 to create the first managed retreat in Essex. Twenty-one hectares of agricultural land were given back to the sea, and have now become saltings. The old sea wall near by is crumbling, the clay cracked with white ragstone exposed. The sea has already ebbed from the older marsh, leaving blue mud, silver mud, brown and green mud. But here the water is still rushing out into the dyke. On the hills are houses and churches all on the brim of this bowl. Against the low sun the Tollesbury lightship, along with boatyard cranes and yacht masts; across the marsh is Mersea. Another angry herring gull circles. A chill wind bites my face, begins to freeze my fingers. A dazzling egret rises, recently rare and now commonplace, blanched alabaster white like nothing else on these marshes. Ligger boards wrapped in chicken wire lead off into the saltings offering mystery tours. Now the wind rises a notch, and yellow cordgrass shivers. A fleet in front of me crumples. It seems cold today but it is nothing like 1963, when the sea froze along 7 miles of the Blackwater. Those were times of persistent cold, deep snows, numbing air, leaden skies and hungry birds, when J. A. Baker was in search of his peregrines. It seems it will never be properly winter-cold like that again. I pour hot coffee from the flask, and make a call to the reserve warden.

Tollesbury Wick reserve is 600 acres of coastal grazing marsh, sea walls and saltings, and an illustration of how coastal farming and grazing livestock can create a fine environment for wildlife. These marshes were reclaimed from the sea in the 1700s when 6 miles of sea wall were completed, leaving a snaking borrow dyke inside the wall. Access to this Essex Wildlife Trust site is normally only around the wall because of the shy grazing animals and ground-nesting birds, but today we walk across the open blustery fields, past wide blue fleets fringed with reeds and echoing with the contented kronks of Brents, over the mid-reserve protective counter-wall, and to the distant rough grazing facing Great and Little Cob Islands and West Mersea. Mike Sandison retired from Ford's after thirty-three years' service, has shepherd's crook in hand, and strides at a pace normal for those who look after animals and always have something pressing to do.

There's a little bit of Scotland on these Essex marshes: 600 Shetland and Ronaldsay sheep and sixty Shetland cattle. All three are endangered breeds and ideally suited to a rough life. We stand in a herd of tough and tiny Ronnies, just 25 kilograms live weight each and thus ideal as they don't poach the often wet grasslands. They are friendly, hobbling as if on high heels to nose around . Slow animal growth is no bad thing for those who farm for conservation rather than just for meat or milk. Mike knows many of the older sheep by name, especially those that have had dramatic lambing episodes. In one of the very first, a ewe with lamb's head protruding leapt a ditch and charged around the reserve for hours before Mike and his wife finally cornered her, and helped deliver two healthy lambs. They were raised on the bottle, and Bess is still here, grazing contentedly over in a small group. You can't get too close to livestock though. Twin Porgy went for meat.

Now, though, a new problem: Blue Tongue has come ashore in insects drawn north from the Mediterranean by warmer weather. No one had heard a whisper of it before, but suddenly it's here to stay. The animals are not for the moment allowed to be moved off the reserve, and Mike will have to limit next year's crop of lambs. There could easily be too little food for all the stock. Mike's now proud chairman of the Shetland Cattle Breeding Association. These Scottish migrants have recently moved from critically endangered status to just endangered, partly because conservation organisations have started to use them and other traditional breeds, such as Konik horses and Hech cattle, as management tools. We stand quietly among these short black-and-white cattle, murmuring to individuals as they jostle around us, lowing gently, at home on their Essex marshes.

I walk around this grazing marsh, beginning where the embankment of the Crab and Winkle railway line sweeps out of woodland and heads for the old terminal by the pier. The Kelvedon to Tollesbury line was opened in 1904, and the pier soon after with high hopes that Tollesbury would be the next Clacton or Southend. There were four trains daily, and passengers could hail the train anywhere along the line. But in the year it opened, there was tragedy as 12-year old Frank Leavett was killed after he sat by the line and fell asleep. The resort never took off, and the 700-yard pier was used instead to bring coal ashore. The pier went in 1940, when all those along the coast were cut by the war authorities, and the line then folded in 1952.

I rest at the end of the line by a pillbox tumbled into the Blackwater. Industrial materials, bricks and clinker form the base of the pier, and lines of wooden staithes march into the incandescent water. Behind, the Shetlands are still lowing mournfully, and then are joined by the higher-pitched bull. Ahead a red-sailed ketch glides past, midstream. It is November, but it's no wonder Blue Tongue's vector insects are surviving. A haze of flies and midges swirls around me. Tollesbury Wick is a great prairie of flowing grass, the roughest sectors yellow, the improved green. I am at a point not far in miles from settlements and civilisation, but wild under a pale-blue sky, with a scattering of clouds on the horizon and stretched high cloud ripped sideways by the jetstream. The tide creeps up another metre as I sit and wait. A marsh harrier glides over the grazing marsh; then a peregrine settles on a nearby fence post, swivelling its head slowly from side to side. Skylarks seem strangely indifferent and continue to sing. I pick up a sliver of wood from near the pier, a shard of brick, and a grey-and-rufous falcon stone to add to my collection, and walk towards Shinglehead Point.

A pillbox with a faded CND symbol on the roof looks across to Bradwell power station. I stand on the roof and peer along the Blackwater, and then jump down to walk across the saltings to the beach. On the pillbox faded white graffiti says *Shut Down Bradwell*, and so it was. To the author's surprise, no doubt. Across the channel is jumbled West Mersea and beneath my feet are yellow horned-poppies and mounded hills made by the rare yellow meadow ant. There are brown hares and foxes somewhere here too. Around in Woodrolfe Creek are ranks of moored boats, and the bright-red lightship stands above them as the only saturated colour on these saltings. The wind rises and there is the insistent plink, plink, plink of lanyards against masts, and still the burblings of Brents from the fresh marshes. My fingers are now so numb that it is difficult to write. The sky is cobalt blue, and the cold wind not so strong as to drown the whispering reeds.

The marina is both history and future. There is every kind of active and abandoned boat: some holed and half full of water, others drawn up on saltings, some just bare skeletons, others inverted and waiting an owner's return. Ligger boards cross the marshes, some slick with sediment, and others with planks missing: easy to spot in the daylight, but a trap on a dark night. The boatyard is active, buzzing, the marina chock full. It was built in 1959, and signified the final shift from a fishing economy to one based on leisure yachting. Men from Tollesbury, Wivenhoe, Brightlingsea and Rowhedge raced in the America's Cup from the 1890s to 1930s as helmsmen, captains and crew, and were said to be the hemisphere's best sailors, having learned their trade routinely racing their fishing vessels against the tides to get catches to market. I pause by blue water and yellow marsh, and hear echoing shouts: a tractor draws up ready to haul in a boat. Behind the yard stretches the reserve, wide and wild. And managed.

Richard Mabey and I come back here in the spring to see how things have changed. The bluebells are beginning to show, but it's before the white hedgerow spectacular of cow parsley and hawthorn in blossom. But it's not spring, for another easterly wind zips biting air from over the sea. A lazy wind that goes straight through you rather than around. We walk around Tollesbury Wick Marsh again, clockwise this time. Nothing moves. Birds dare not risk flight in this wind, and we content ourselves with imagining what is there.

"Shall we go for lunch?" suggests Richard.

We repair to the Chequers in Goldhanger to sit in the warmth beneath low dark rafters, and it feels like only a few days since I stayed here last summer. We go in search of the Osea wetroad. Snaking before us, rising out of the water, is the road that twists and jinks across the muds to Osea. It should be low tide, but the wind is holding some of the water in late. We set off past a line of waiting cars and vans, and wonder at the amount of traffic heading for that tiny island. We walk out to sea along the 3-metre wide hard, and pause to watch a flock of black-tailed godwits out on the muds. Oystercatchers pipe and wheel across the flats. Sticks mark the route for those unfortunate enough to have to cross when water covers the road. We get most of the way across before finding that it would mean soaking wet feet to go to the end. The island is still an island, and we turn around with the wind now at our backs.

There's a car in the mud. It's 20 metres out from a pillbox in the sea wall. The roof and windows are gone, and mats of bladderwrack hang from the struts. How many tides has it watched come and go? How did it come here: pushed or driven over the wall, or floated on water? It faces the sea, exhausted, chassis-deep in the mud. It's early winter and I have returned to the Dengie to search for some of the wildness of the coast. I've struck east from Southminster along dark empty roads and into the reclaimed marshland. I climb onto the sea wall and see there is to be no sunrise today, just a gradual lifting of the grey gloom, and the revealing of a flat and open land. Clutches of low trees are scattered across wide fields and water gurgles in ditches. To sea the mud and seaweed stretch out hundreds of metres. A flock of half-hidden waders flits by. An egret lifts from the saltings, crying angrily. These birds join the grey heron, called the frank by old marshmen, in bringing harsh calls to these lands. I walk south towards the mouth of the Crouch where smooth waters hide rip tides and mobile sandbanks. On a corner of saltings are two wooden crosses. I walk out through the cordgrass and sea lavender and find one cross beam is nailed to its stem, the other roughly held in place with barbed wire. Like the car, there's nothing to say why they are here, nor who left them. Across the fields, the turbine roar of a grain dryer starts up, and will become a constant companion for the rest of the morning. Dengie: still a mix of the wild and used.

Further west, near another pillbox slotted into the raised sea wall, looms the unique hulk of the concrete command tower with its eye-slits and pointed roof, wrapped today in mist. The contrast with my summer visit could not be greater. Further inland I see a white van with shining headlights crawl along a track, and turn inland to disappear back into the mist. A small plane passes overhead, invisible too. Some secret mission, perhaps, risking all. Still the grain dryer roars, and now geese join in, brents from over on the Foulness side. I can see the low line of Foulness and its concrete and grass-topped sea wall. There's just a single shrubby tree. According to John Leather, the Light Railway Commissioners granted powers in 1902 for the construction of a railway from Southend to Colchester. Had it been built, it would have changed this place forever. The line was to cross the Crouch here by steam ferry, travel through the Dengie to Bradwell Quay, cross the Blackwater again by steam ferry, and then continue from Mersea to Colchester. Luckily, perhaps, the money was never raised. On the return journey, I notice even more mosquitoes following me along the wall, and come upon the body of a gull, hollowed out by one of the descendants of Baker's peregrines, grey and white feathers scattered on the grass.

The ague was a mostly non-fatal form of malaria, and it stalked these marshes not so long ago. A common greeting as winter ended was to ask, "Have you had your ague this spring?" But there was no need to bother to ask come the autumn, when malaria peaked again, as almost everyone suffered attacks. In historian John Norden's 1594 trip through Essex, he reported being bitten "especially near the sea coastes … and other lowe places about the creekes, which gave me a most cruel quarterne fever". Also called marsh fever, it produced sallow sickly faces in sufferers, and swollen stomachs, particularly in children. It was the source of one great myth of Essex that has been widely and uncritically quoted. Daniel Defoe, not altogether given to double-checking everything he heard, and who seemed to write about places he had never been to, wrote that the Essex men of the marshes and islands would commonly take five or six to fourteen or fifteen wives. They fetched them from the inland uplands, and when each died of the ague, would return for another. It's got to be exaggerated nonsense. You can imagine: an old boy spins Defoe a good story in return for another pint or two. There would be some truth about local immunity and what they do drink is beer laced with opium obtained from local poppies, served at the time in all the pubs of Essex and Kent. But fifteen wives?

Tollesbury lightship

It's been gone a century, but it's not clear why: a mix of drainage, the increased use of quinine for treatment, and the discovery of the role of the *Anopheles* mosquito in its transmission all came towards the end of the 1800s. Marshmen believed that gold earrings warded off the ague, and also gave them long sight. But in 1904, Herbert Tomkins was still reporting that the ague was "in most every family", forcing sufferers to crouch by the fire, constantly shivering or sweating. The last officially recorded outbreak of malaria in this country occurred in 1918 when injured soldiers evacuated from Greece brought back malarial parasites in their blood. They were taken to the Kent side of the Thames at Cliffe, and 500 local people caught malaria vectored by local mosquitoes from the soldiers. Although malaria has not yet returned, it does seem that mosquito numbers are on the increase. There are thirty-three species in the UK, five of which are of *Anopheles.* Autumn 2007 saw huge numbers of mosquitoes in marsh country, probably because the summer was so wet, leaving many stagnant pools of water. Some people on the marshes have started sleeping with nets. Maybe it won't just be malaria we need to worry about. The Asian tiger mosquito, a carrier of dengue fever, encephalitis and yellow fever, is now established in Italy, and was recently found in Belgium. If malaria were to return to Britain, it would almost certainly be to these marshlands of the east. I am slapping at my exposed skin, but miss many. Angry bites stay with me for days.

Mosquitoes may be common but over the river it is the badgers and foxes that are a wildlife mystery. Anne Chittock has invited us to walk the top and west of the military island. She was born and grew up on Foulness, went away, came back, and became chair of the parish council. Once an islander, always an islander, she says. Gill and I sign in at the security gates and time-travel up the long spinal road, past solitary military-brick buildings with metal windows and flat roofs, cranes for testing weapons, a group of golf-ball structures and side roads blocked with red barriers. There's not a single person in sight. Foulness islanders are a people apart, with little connection to the mainland, and it seems it will stay that way: the residents campaign for Broadband, but the military say no. Anne's husband, John, used to be manager of local operations. The largest ordinance tested are giant 16-inch shells, fired through targets on the northern tip of the island, landing on the sands and mudflats beyond. Near by, the booms are not so bad, but further along the coast people feel the shaking more. "It's not so much that the house is shaking," John explains. "It's the ultra-low frequency making your body shake. The liquid in your eyes wobbles."

At the north-east of the island, brents are feeding on the dark eelgrass, crowded by the rising tide. When the *zos* seagrass almost disappeared due to disease in the mid-twentieth century, the numbers of Brents fell sharply. Now they are on the up. On the inside of the sea wall are lines of concrete defences built as replicas of the German defences in Normandy, and for the troops to practise on before the D-Day landings. Further along is a copy of the Japanese defences in Singapore. Out to sea are the distant forts at Shivering Sands and Ray Sands, marching across the horizon like invaders invented by H. G. Wells. Designed by Guy Maunsell to provide a defensive ring for the Thames, there are three outer naval sea forts at Sunk Head, Knock John and Fort Roughs, and inner Army forts on stilts at the mouth of the Thames. We pass outward-facing signs warning of danger, no entry, designed to repel maritime visitors. Clouds of golden plover twist and turn, and are joined by smaller jinking groups of lapwing, black-and-white against the low grey clouds. We search the sky. If there is a hawk or falcon, we can't see it. Cormorants are working the river, suggesting a sprat run. We look across to the distant mine tower, where a day earlier I had looked this way.

There are seven badger setts on the island, and lots of foxes too, but they're all up here at the northern tip. Stilt-legged avocets have recently begun to nest on scrapes by the borrow dyke, but if the water dries out, then both badgers and foxes have their eggs and chicks. One sett was found in the sea wall, and had to be removed for fear that the seawater would find an easy way in. But their presence remains unexplained. Why did these two species not just stop as soon as they had crossed the southern bridge? It looks like they were deliberately released on these northern shores from the sea.

By the only tree, a gnarled pear, is a concrete hard, site of the cottage destroyed when the walls first gave way under the force of the great tide of 1953. Anne remembers waking that morning, and thinking there was snow outside. Her mother had heard a ticking noise in the night: it was a floating table knocking on the ceiling. Tides have not since come over, but then the walls are much higher. Inside a dark marsh harrier rises out of the reeds, and silently drifts around in a wide circle. Partridges scamper ahead of us, and we can see a shoot walking out on the farmland. From Nase Point we follow the Roach, and turn inland from the quay where a couple of months earlier we were ferried across to Wallasea. Today is quiet, even an elderly birdwatcher pointedly looks away and says nothing when we offer greetings. We head back for lunch and gaze out of the kitchen towards the elevated sails stealing up the creek that marks the edges of these islands of the archipelago.

COLCHESTER
HYTHE BRIDGE
N
W
E
S
University of Essex
OLD HEATH
Hythe Mashes
WIVENHOE
ROWHEDGE
ROMAN RIVER
FINGRINGHOE
Geedon marshes
MERSEA ISLAND

Chapter 6

ESSEX

Chapter 6
Wild by Industrial

In the warm darkness, a tawny owl hoots, and the heady scent of lilac fills the air. I go in search of angels of the night. The town is quiet as I pass the walled garrison, remnant of occupation 2,000 years ago, the darkened pubs, British Grenadier and Royal Mortar, and the great clapboard military church. I navigate south to the nature reserve perched on a rise above the firing ranges that stretch south to the sea. As light inches over the horizon, I fiddle at the lock on the gate, and let myself into the mixed woodland that has sprung up over the forty years since the sand and gravel works left this place as barren as the moon. The Essex Wildlife Trust acquired Fingringhoe Wick as its first reserve, and has made a success of this place. The geese that have accompanied my walks during autumn and winter have flown, but in their place have come shy nightingales, and there is magic in the air. In recent years, these rufous-and-white winter residents of West Africa have been flying north in growing numbers during spring to skulk in deep thickets and shrubs, serenading female *Luscinias* and charming anyone close enough to hear their breathtaking song. Across the country, deer have been no friends of nightingales, as they clear woodland of low scrub, and nor has the recent decline in proper coppice management helped. But here, a thick understorey hides dozens of them, and they're arguing, competing, pausing and soaring in song. I stand in wonder, walk on, stop and listen to another, and move on again. Here they are, in almost every other bush, it seems.

Nightingale as a name has been used for at least 1,500 years, and the Saxons of the region knew the bird as the night songstress. But it is only the males who sing, and they will do so in the most remote as well as urban places, perhaps competing with traffic noise. They'll sing all day and night, but are most melodious at dawn. It's the national bird of Bangladesh and Iran, has been the subject of poets and folk singers of many countries, and plays a prominent role in our own culture, perhaps partly because of its scarcity. They never seem to cross north of a line from Norfolk to Hampshire, and number at most 400 pairs. Yet we've all heard about them, even though most of us have never seen one. We know about their rich liquid song, mixed with gurgles, whistles, drums, chuckles and compelling breathless passages. Richard Mabey wryly comments, "It is a performance made up, very often, more of silence than utterance." This morning, I may be wrong, but when these nightingales sing, no other bird can possibly compare. In 1940, Eric Maschwitz wrote a wartime song with Sherwin and Strachey about a nightingale singing in Berkeley Square, and it somehow captured the romance of summer nights, and the hopes that the war would come to an end. It was first performed by Judy Campbell, became a standard for Nat King Cole and Glenn Miller, and was probably most famously sung by Vera Lynn, who sang of other birds too, blue ones, over the white cliffs of Dover.

The eastern sky now grows indigo and then tender pink, and reflects in the wide Colne River that gently laps the shore. A mist of mosquitoes surrounds me, even when I sit in a hide to gaze across to distant oyster creeks past Geedon, where Wentworth Day saw two sea eagles "swinging in lordly arcs" in 1931. A fox barks, then again fifteen times, and a male cuckoo calls repeatedly at the dawn. Blackbirds and wistful robins join with silky songs; a pheasant clatters harshly by. But all the while, there is the rolling, insistent rising and falling, the "churr-churr", a buzzing interlude, silence, and then they're off again. I do see one brown bird in a bush, but even though I look hard none of the others reveals themselves. They seem to have the power to remain invisible. I sit at a bench to drink coffee, and look down to the now silvery-grey water, and the fox is barking on and on, forty, fifty times. And the damn mosquitoes have drawn blood more than once. Matt Cole, the warden at the Wick, either has nets in his home or else he and his young family are made of seriously tough stuff. The warm green oaks appear from the gloom along with bright hawthorn and their white flowers dripping the scent of spring. The broom are bright yellow, but the golds of gorse have faded and their best days are over. It is also bluebell time, but not here. Bluebells and nightingales don't get on. The one needs open mature woodland; the other thick scrub. On this day, I will count thirty-nine species of plants in flower, and another ten grasses and sedges. A month ago, there were only five. This spring advanced in February as winter faltered, receded with snow at Easter, and then in the week before this walk there was constant sunshine and temperatures typical of July and August rather than May. The old weather patterns seem disturbed. We might have to get used to this.

The sun climbs over distant Alresford Creek and I leave the reserve by the sea

River Colne

wall. Half a dozen cuckoos are now in active competition, and I slip past private signs to head north-west towards the gravel works at the appropriately named Ballast Quay. The deep grass is damp with dew, and ahead of me I can see a canopy of white cow parsley twisting in a line to mark the transition from farmland to saltings. My shoes are soon soaking, but now the sun brightens the exposed muds. The ribbed skeleton of a barge appears, and then I gradually become aware of a new sound. The exposed muds are fizzing as the tide ebbs. I look carefully, but cannot see anything more than water forming into trickles, and then longer sinuous curves of rivulets, going on to join the main river. The mudbanks are wide and dangerously deep, yet there's no visible sign to account for this bubbly effervescence. Champagne river, this Colne.

The shores of the river, once a main Roman route, are chock full of gravels and sands. And now alongside the wild is the distinctly industrial. First nightingales and foxes, now a blasted and extracted land. There are tailing ponds of soft sands and silty muds, and great piles of graded gravels. Yellow dumper trucks are the size of houses, and cavernous barges tied alongside the quay await loading. At this time

Rowhedge Quay

of the morning, all is peaceful. I walk up a truck route into the forest, but as time passes I notice that the sun has to be coming from the wrong part of the sky. I go back, dodge under conveyor belts and find a gate to the village road. Bedroom windows are flung open to the cooler night air, and mosquitoes too, I imagine. It's still early, a little after six, and the village is sleeping. Cats prowl, and I am walking, dew-wet shoes creaking on the tarmac.

I continue inland to cross Roman River at the tide mill. A colony of rooks is animating the morning air with glorious chatter. I think of the incomer over on the next valley to the north, the Stour, who bought a big new house by an ancient rookery, and took a gun to the birds because he couldn't bear the noise. The birds left and never came back. Who said we humans aren't in trouble? But all is quiet on the people front today, and there is still no one abroad on this magic morning. I walk across East Donyland Marsh, through a pathway under a dense thicket of grubby woodland, and come out at the back of desolate Rowhedge Quay, another industrial wasteland waiting for a new vision. These days it will almost certainly involve housing. We are replacing what were once manufacturing and trading facilities with more places for people to live. Once barges and ships brought goods in and out, fairly close to where they were needed. No longer. Now trade is centralised and lorries on roads do the hauling. It's changing the nature of this coast.

At the wharf is a line of a dozen high-bay warehouses and an expanse of cracked concrete gradually being split open by forceful weeds. A white sign, roughly painted, says *Do not open this door*, and great tyres are piled in front. Surely some local lads have been tempted. Everywhere is wire fencing and piles of concrete blocks. It's a purgatorial landscape, neither one thing nor the other. I arrive in Rowhedge, former affiliate of the Cinque ports, and pass two riverside pubs and the mix of new waterfront vernacular and tiny cottages that presumably once housed quay workers and fishermen. All those livelihoods mostly gone. Here the Colne is so narrow you feel you could jump across to Wivenhoe. It's now the smallest of all the arteries on the coast to north Norfolk: from Thames, Crouch, Blackwater, Stour, Orwell, Alde/Ore, Blythe, Waveney/Yare, and finally to the Great River Ouze. The Colne was wide once, it's just been allowed to silt up.

Two and a half hours of walking and the first person of the day appears. She smiles and nods. Her two dogs simply walk by too. Around the bend boats are clutched by thick, slick mud. Two greylags wander across the sculptured banks, and now cropped-grass river wall marks the start of Hythe Marshes. To the left lie raised lagoons, unmarked on my map, and something to do with the nearby sewage works that can fill this valley with offensive odours. On the far side of the river is the university on the hill with its six tall brick towers, proudly built in the mid-1960s, and dominating the skyline of the campus that once was a deer park. On the north side runs another riverside railway line. And now after these green marshes, another transition, back to more post-industrialised landscape on the southern edge of Colchester, a town with so much history yet like many others suffering at the hands of those who thought inner-ring roads, prefab blocks and car parks were all improvements.

Colchester is a town of antiquity and continued settlement. It nearly became the country's capital too. It was founded by the Trinovantes, who built a 16-square kilometre oppida on a hill, and called it Camulodunum, the fortified place of Cumulos, Celtic god of war. It had a royal mint, sacred burial sites, and fortified dykes. By AD40, Cunobelin was King, and had united all of south-east Briton under him. But three years later, the Romans came: four of the Empire's twenty-seven legions invaded. At the Thames, they waited for Emperor Claudius, and with members of the Praetorian Guard, elephants and several senators, swept up to Camulodunum.

There was no choice. The town surrendered. The Roman Senate saw the significance and rewarded Claudius with the title Britannicus, and by AD 50 the capital was further accorded the highest rank of Roman cities as a Colonia. Its entire population was thus granted citizenship of the Empire. The Temple of Claudius was built, along with a 3,000-seat theatre, monumental arch and a fine market, but defence was ignored. Tacitus recorded that the native people thrown out of their homes were unhappy, and that the Temple was seen as a potent symbol of colonisation. The good times came to an end. In AD 61, Boudicca, Queen of the Iceni, razed the Colonia, slaughtered citizens in the sacred groves, and left evidence seen today in the centre of modern Colchester as a burnt stratum, known as the Boudiccan destruction layer. The Romans came back, and after more bloody battles rebuilt the temple and established the first city wall in Britain, nearly 3 kilometres long. Then in AD 80 it was over. The administrative capital was moved to the much smaller Londinium. And thus the face of Britain changed.

For nearly two millennia the town was a seaport, but you'd hardly know it today. Up on the high street above the river, the town hall has cornerstone figures representing four elements of the local economy: engineering, fishing, agriculture and the military. Only the Army remains as a significant presence. There were port facilities at Old Heath from pre-Roman times, the name morphing into Old Hythe by the twelfth century. But this river was always silting up, and every century was marked by desperate efforts to improve the navigations by forced removal of weirs and fish traps, by dredging, and by projects to recut the channel. By the early 1600s, the channel was impassable between Rowhedge and the Hythe, and six national Acts for improvements were granted between 1623 and 1818. Commissioners were appointed, but sat on their hands. In 1842, the navigation committee rejected a plan from Peter Bruff for a new dock and cut, yet by the 1880s, Colchester traders were complaining that it took ships five to seven days to reach Colchester from Wivenhoe. Steam dredgers then reopened the river and soon Thames barges could get up and down. In 1900 King Edward Quay was built, and four years later Herbert Tompkins wrote that "on the quayside were warehouses and coalyards; the rattle of machinery rushes out upon you from the great oil-mills; cranes clank incessantly as the barges are loaded with oil cake or with casks of linseed oil". The quay was extended in 1925, and levels of river traffic further increased. But then no more. Though propositions for further improvements would continue to be made and rejected, the port was closed in the 1990s. Silt had won, and so had the roads.

The quay is completely abandoned. All is quiet. Empty warehouses look as though they've been untouched for years. Graffiti brightens walls, but there are also piles of fly-tipped rubbish, magazines and broken videos. There were once famed boat-builders here, now the only building is of new blocks of flats. How will they look in twenty years' time? Next to succumb to the housing boom will be Gas Quay, where once coal from Newcastle was converted into gas to light the town. The works are long gone now, but were flooded that night of the great tide in 1953, plunging all of Colchester into darkness. I find old fertilizer stores, grain silos, an electricity sub-station, metal recycling facilities, scrap merchants. All rust and crush, an industrial wilderness, a desperate place. A couple of months later, a man stepped off this quay into the deep muds. It's an enduring nightmare of wildfowlers and oystermen: to be stuck in the silt until the rising tide snuffs out your last breath and the crabs take your eyes. He was stuck fast up to his neck for four hours before being discovered by workmen just as the tide reached him. Only power hoses could free him for the emergency run to the hospital. No one knows how mud-man came to be there.

At the Hythe bridge I start back down the river's north shore. The trail below the university is populated with huffing runners with dogs and speeding cyclists. It's the busiest patch of the river today, and makes up for hours of earlier silence. I walk past a row of rare black poplars, and then a rich patch of woodland and acid heath just renamed Nightingale Wood. Just outside Wivenhoe, another nightingale sings from the bushes. A young woman with a baby in a pram stops, and we stand and listen. Three runners pass, each wearing white headphones, missing this music of the woods. We don't say much, even the baby is quiet. On the edge of the town, I come to the deserted railway station with its echoing announcements, and pass through empty streets back to the river. On the far side the Ballast Quay works are now clanking with activity even on this Saturday morning. Ballast was an enormous problem for harbour authorities: empty ships needed it, those waiting to pick up goods had to jettison it. On this side Wivenhoe too has fallen under the spell of housing development, and on its southern reaches are tightly packed new homes by the river. Richard Rogers says these kinds of modern developments will become our future slums, with too few parking places and building materials prone to ready degradation. But the official sign says *Waterside living at its best.* There has been recent controversy here too, as health-and-safety spoilsports had become worried about youngsters jumping off the bank into the river a few feet below. A group of residents took away the ladder which swimmers used to climb out of the

Richard Haward, oysterman

river. Some have got this outdoors thing all wrong.

Further on is the Colne's giant barrier, built in the early 1990s to prevent flooding from storm surges. Yet this was before concerns about climate change emerged. More additions to 1,000 years of interference on this river. I am back to the wider reaches of the river where green inland marshes are scattered with the pure gold of buttercups, and hedgerows are lined with oaks of delicate green and more blossom of hawthorns. A metal memorial bench is dedicated to a man *who sailed these waters for fifty years.* Past Marsh Farm with its wailing sheep, lambs following ewes as they walk up the hill, bleating too for all the marshlands, and then I have to head inland again to negotiate another tributary. The map says there is a ford across Alresford Creek. Not a chance. There are great drifts of mud, and although the tide has now receded and the water is only a couple of metres wide, only a fool would try to cross. Boats are tethered by the saltings, some lived in permanently, smoke issuing from small chimneys, bicycles propped on the shore.

Here's another pit with its sand and gravel works. In the creek are old quays and rusting architecture. More signs say *Keep off.* On the far side on the 20-metre contour is Brightlingsea's flint church, looking down on both this creek and the main Colne River. Oddly, the town is down on the next creek, and the church almost alone on the sole route into this river's other Cinque port. I walk on past farmland and ancient oaks surrounded by stands of cow parsley. In a dusty brown field where young cereal shoots are pushing up are lapwings, invisible until they leap into the air, crying as they rise, twist and fall, distracting intruders away from their nests. I cross at the mill, passing a pond thick with duckweed and downstream swaying reedbeds, and then walk up the hill to the main road, where broken car lights are scattered in the gutter, this road so often closed by careless drivers taking too many risks. I press on up to the knapped flint church on the hill, one of the grandest in all of Essex, and where you can see across to Baring-Gould's island of Mersea and its hinterland of marshes and creeks full of oysters, where the sea is wet as wet could be, and the sands dry as dry. I wait by the churchyard, and then a taxi whisks me back to the Wick, where Matt is digging and whistling in the vegetable garden, a film of sweat and soil covering his skin.

"Good walk?," he asks, and turns another sod.

The day is still to get properly hot, and there is much to do here.

Oysters are Colchester. They have been raised in the Colne's estuaries and creeks for at least 2,000 years. It's a long time for any kind of human activity, let alone one that has hardly changed over the centuries. The Romans loved oysters, and for the past 800 years Colchester has celebrated them with an annual feast in the autumn. Richard I granted the Corporation of Colchester a Charter in 1189 that confirmed rights to all the oysters of the Colne and its creeks. Oysters became part of national culture and were not always exclusive fare. Dickens has coachman Sam in *Pickwick Papers* comment that it is "a wery remarkable circumstance, sir, that poverty and oysters always seems to go together", and Lewis Carroll's liminal and upside-down *Alice Through the Looking Glass* makes oysters victims of the walrus and the callous carpenter. The two companions persuade young oysters to hurry up, all eager for a treat, and then tell of many things, of shoes, and ships, and sealing wax, of cabbages and kings. Out comes the bread, pepper and vinegar besides. The butter's spread too thick, says the carpenter. But the walrus to the oysters said, I weep for you, I deeply sympathise, and with sobs and tears, he sorts out those of largest size.

The wind whistles off the marshes, and the clinker-built green-and-white oyster boat circles in the creek. The winch clatters as the dredge is hauled up and released, and oysters cascade across the working bench on the boat's port side. Now the management begins. I had joined silver-bearded Richard Haward at first light on a February morning at his family's famed Company Shed down on the Mersea front amongst boatyards and sheds. The fishmonger's and restaurant was run by his wife, now by his daughter, and soon contains a sparkling display of fish and shellfish. The smell of the fish shop is one of three that launch me back into the memories of growing up in a seaside town, the others being the malt of brewery and the freshness of new bread. Though the Company Shed has neither booking system nor licence, there's regularly a queue of people waiting for a table. It could be bigger, but why take on the hassle, says Richard. Here small really is beautiful.

His part of the operation centres on the oysters. They are actively managed, though with far less control over conditions than any farmer would accept. The quality of water is vital. *E. coli* bacteria from human wastes and wild birds can contaminate waterways: only if waters are graded A can oysters be harvested and sold without treatment. Grade B means the shellfish have to be treated with ultraviolet light before consumption, and grade C oysters have to be relayed into A or B waters before harvest. It's tough. There is also the dreaded bonamia parasite and occasionally severe pollutants. The uncertainty is constant.

Oysters are raised on platforms by the edge of the creeks and fleets, or in the channels where they are dredged. Richard and his son also sort out those of largest size, dropping them into a plastic box; clumped groups are split apart with knives and put back in the water, and others have predators prised off. They return the shells to the estuary bed to encourage the spat to settle, but the slipper limpets are kept and tipped out on great middens on Cobmarsh Island. No return to waterways for these pests introduced from the US in 1880. Gulls scream and cry, wheel, rise, fall, and then float motionless on the wind. Why do gulls get such bad press, and yet terns are seen as delicate and worthy? In the tub is now a mix of native oysters and the larger Portuguese, now called gigas. When Hector Bolitho joined men working a boat in 1929, he said the work was "cold, slow and disappointing". But when he was given a Colchester and an Iberian oyster to compare, he said, "The Spaniard was a trifle bitter, the Colchester was rich and pleasant."

Some say the Romans brought oysters to the Colne, but in 50 BC Sallust admitted they were here already, saying condescendingly, "Poor Britons – there is some good in them after all – they produce an oyster." Yet at no time in recent history have there been so few boats working these waters as today. In the mid-1800s, the Colne Fishery Company alone had 500 oyster smacks and employed more than 2,000 men. There were other layings on the Blackwater for Tollesbury, Maldon and Mersea fishermen, and also on the Crouch, Roach, Stour, Walton Backwaters, Deben and Alde. Inshore fishermen bring benefits to their local communities in the form of identity, place and history, as well as income. Yet from Leigh's cockle fishers to King's Lynn on the Wash, there are only a few boats remaining at each port. Regulators have limited fish catches with strict quotas, which means by-catch is thrown back dead or landed illegally, or have decommissioned boats, causing whole livelihoods to be given up.

But there are still fish to be caught. Last summer brought a glut of brown shrimps off Lynn, and men travelled from across the region to work those boats, earning up to £1,000 per week. Yet the shrimps were sent to North Africa to be shelled and packed, and then brought back to northern Europe for sale. Richard is seeking Slow Food praesidium status for the fishery here, and hopes this will help secure his future and those of the oysters. Somehow the behaviour of consumers has to change if those who produce local and wild foods are to survive, and then to pass on their traditions. What happens, he says, if no one wants to do this any more? He climbs gingerly out of the boat, and we lug the oyster boxes back to the Shed. Later, as I walk away, a group of small children rushes down the pontoon, crab lines in hand, buckets and chunks of bacon at the ready, opportunity beckoning. The world's their oyster, so said Pistol to Falstaff.

N
W
E
S
SUFFOLK
SUDBURY
STOKE-BY-NAYLAND
Tiger Hill
EAST BERGHOLT
BURES
NAYLAND
DEDHAM
WORMINGFORD
MANNINGT
ESSEX

Chapter 7

ORWELL BRIDGE
NACTON
RIVER ORWELL
Pin Mill
Shotley Marsh
ERWARTON
SHOTLEY GATE
FELIXSTOWE
HOLBROOK BAY
HARWICH
RIVER STOUR
MISTLEY
WRABNESS
DOVERCOURT
LITTLE OAKLEY
Crabknowe Spit
HAMFORD WATER

Black poplar on the Stour

Chapter 7
Artery & Estuary

The waters are crystal clear, and we are flying over a jungle. On the surface are great cordate lily pads with both yellow and white flowers, and the purple and pale petals of arrowhead are pushing out of the canopy into space above. Below is a mysterious rainforest of sprawling trees with swaying leaves, spikes of rushes, gossamer grasses and pikey deeps. We glimpse beasts darting between the trees, fish shadows amongst the vegetation. We glide on. Roger Deakin once canoed down the Waveney with Norfolk on his left and Suffolk to his right. Here we have Suffolk to port and Essex to starboard. Your eye is at the same level as the fields, and being so low brings a different perspective to the land. There are hidden places, wet glades in woodland, water rushing over weirs, silent pools with swirling midges, shoreline tracks of invisible wild animals, and great alders, black and match poplars, willows, ash and oaks. You become "just a floating being", wrote Deakin. "Your self goes somewhere else." This river is a meditation, as it quietly flows through glacial valleys to the sea.

The Stour rises in eastern Cambridgeshire, and then for most of its course defines the border between the old kingdoms of the Angles and Saxons. It was the first river in the region to be opened up for barge navigation in 1705, and is best known in its lower reaches for being Constable's country. One summer's day, we start at fellow painter Gainsborough's home town of Sudbury to paddle a modest 24 miles to the estuary at Manningtree. After a century and a half as the inland motorways of trade, rivers and canals fell into disrepair after the advent of the railways, and were later comprehensively eclipsed by the internal combustion engine. Barges on this river carried 12,000 tonnes of coal annually up to Sudbury, green transport that would require 250 of today's largest lorries to do the same. Those twenty-seven foot barges drew only 3 feet, took fourteen hours to make the journey upriver, and then two hours less back down. Sudbury was a famous weaving centre for silks and satins, and made the Queen's coronation robe in 1953.

Now gentle waters wind and twist away from this former manufacturing town through lands mostly unseen and often forgotten.

Gill and I launch a Canadian canoe from the dock by the theatre, and glide across a glassy pool overhung with dark alders and out into Cornard Mere. In minutes, the silence of the river advances. Or rather, traffic sounds recede. Now comes a new soundscape of songbirds. The river is green and deep as it lazily meanders around the grazing marshes, the mirrored surface broken only by our paddles. Insects catch fire in the low sunlight. An empty moorhen's nest sits high on a platform of woven reeds. Here the valley widens, and herds of curious cattle crowd against the water's edge, gazing intently, chewing silently as we glide past. There are only a couple of working locks remaining on the whole river, the rest replaced by concrete weirs. At one portage, a magnificent black poplar throws out improbably angled branches, the first of several specimens along this river but a national rarity.

We're coming to the land of dragons. Before Bures, though, are the acid grasslands of Tiger Hill, where the tusks of a sabre-toothed tiger were found, and the glades, flowering banks and rides of Arger Fen's bluebell woods, where Ronald Blythe and friends meet for a celebratory picnic each spring. Also over to our left is a rounded hill topped by a simple stone chapel. The 15-year-old Martyr King of the East Anglians, Edmund, was crowned up there in AD 855, and reigned for just fifteen more years till he met a grisly end, and was, well, buried at Bury. Bures was the location of one of the only two dragon myths over 1,000 years of history in East Anglia, the other at Bulphan Fen in south Essex. In 1405, John of Trokelowe and Henry de Blandforde reported that a dragon had come to the territory near Bures: *vastus corpore, cristato capite, dente serrate, cauda protensa, nimica longitudine*, they said: vast in body, with a tufted head, saw-like teeth and a tail immeasurably long. The dragon was said to have killed sheep, and put the countryside into turmoil. The bowmen of Richard of Waldegrave were sent to kill it, but their arrows sprang back from its armoured body. It eventually took refuge in a deep mere downstream at Wormingford, where, some say, it still resides. Ronald, though, has another story. The dragon was a crocodile brought from Egypt to London, which then escaped and swam down the Thames and up the coast, ending up in this river. In China, the crocodile is still today called a dragon.

Leaning limbo low to slip under a bridge built with only road navigation in mind, we drift into a stretch of the river edged with flote grasses and yellow phragmites reeds, with dense clumps of blue-grey rushes in the middle way. Now it becomes

Brent geese on the Stour

narrow and enclosed, and we pass through a tunnel of towering vegetation. Afterwards everyone says it was like the Amazon, even though none has actually been there. For a while, we were lost in an alien place. The sun shone up from the dazzling still water, and then we found ourselves paddling behind a family of swans, two marble adults and three large grey cygnets. We were all in a line, sailing through the grasses, reeds and rushes. We felt quite alone, and could have gone on forever.

The river opens up again, past flanking hills and on to Wormingford. Scattered farmhouses are high on the valley slopes. To the south in a wooded valley nestles Ronald's sixteenth-century longhouse, previously home to John and Christine Nash. To the left, the simple Norman church of St Mary's at Wiston, its cream interior decorated with early medieval wall paintings and lit from small windows up by the rafters. This is where another dragon crouches, for high on the limewash opposite the entrance is an earthy red-brown serpent, looking back over its coiled body, a great beak open and with pointed horn. The strange thing is that the frescoes are dated to the thirteenth century. Up on the hill behind the church is Nayland's former sanatorium, established in 1901 by Doctors Jane Walker and Grace Griffith, where consumptives lay outdoors in beds on rollers or sat in bathchairs for the view across this valley.

We may not be sure about dragons, but we can be about otters. They have come back. By a mill portage, hidden in deep ground vegetation, is an otter holt. Today, there are crimson dragonflies and smaller neon damsels flitting from sun to shadow, and then the blue streak of kingfisher, and we know the river is healthy. Otters were once hunted in this same river by huntsmen in serge suits, but now they are protected and celebrated. A year or so earlier, I had stood one cold winter morning watching a blood-red sun rise out of mist on the river, waiting and waiting, until eventually an otter blurred into the water from the holt's entrance, and left a chain of bubbles as it set off to fish for eels. In summer, you are even less likely to see one of these mysterious animals, even though the Stour has more spraints and other signs of otter activity than any other river in the region. The Colne and Roman River, as well as the upper reaches of the Blackwater, are also well populated. But there are no records yet of otters in southern Essex or on the Dengie or in Tendring. The main threats are illegal crayfish nets with no otter guards and, of course, road traffic. These retreats and advances are a common pattern of natural history. Today, minks are spreading. A century ago it was the grey squirrel that arrived to out-compete the native reds. Writing in the late 1950s, Ambrose Waller said that "the red squirrel is still fairly common, and I am glad to say that his grey brother has not yet penetrated the confines of the Stour". It all changed so quickly. Today there are no reds in any of East Anglia.

Along the riverbanks we see the sad signs of another tree pathogen that has been invading this land. Phytophera now threatens many of the great alders that live with roots in the riverbank. Since Constable's iconic paintings came to shape so much about this region's rural identity, perhaps even England's too, there have been many changes in local ecology. The riverbanks were then mostly clear of trees, so that the horses towing the barges could easily pass. Since then, strings of grey-green cricket-bat willows, also used for artificial limbs, have become common, and copses of match poplars were planted and then abandoned when cheap wood was imported from India. The great elms are gone too. Sadly it may be the turn of the alders. But then there are the black poplars. It's a forgotten wetland tree, a poor breeder, and always a real beauty. You can hear the sonorous rustle of the leaves before you see the tree, distinctive with its branches and trunks that are fissured and covered in bosses. Efforts are being made to plant more along the valley. Ecosystems here are in constant change.

Canoeing these days means regular portages, and the second day begins with more hauling at our Tudor wool-village of Nayland. I point to the grassy bank where a couple of years earlier I had watched a stoat hypnotise a rabbit. I saw it first chase the rabbit across the road, the two jinking synchronously; the stoat dived across in front of the rabbit and rose up on its hind legs and began a sinuous dance, black tip of its tail waving. The rabbit stood stock still, the stoat leapt, and the rabbit flipped away. Again the stoat overtook, danced again, and this time jumped on its prey's back to bite deep into its neck, pulling it into the long grass.

Today we glide around moated Court Knoll, now just an open field, but mentioned in the Domesday Book, and where achaeologists have found prehistoric and Roman pottery, roof and flooring tiles, brick, fired clay, stone, shells, clay pipes, lava querns and animal bones. Still its complete history is not known. We pull the canoes through stony shallows and then gingerly pass the point where a doodlebug was dredged from the river. It's now on loan to the Flying Bomb Museum in Southend. Back in the village is a memorial bench for Lt. Charles F. Gumm, American fighter pilot, remembered for heroics in 1944 when bringing down his damaged Mustang on the fields rather than the houses, and perishing in the crash.

We bend our backs and paddle around The Fen, 30 hectares of common grassland fringed with more grey-green willows, still parcelled into 180 fennages which confer common rights to graze cattle. Looking back towards the village, there is the church looking down on this patch of land that has remained unchanged for at least 1,000 years. Nayland grew on local wool and weaving skills brought by the Flemish, and into the twentieth century still had almost every craft, artisan and shop – butcher, baker, beer-brewer, dairyman, a bus service, builder, cycle shop, fish-monger, farmer, shepherd, chandler, blacksmith, slaughterer, tailor, iron-monger, grocer, draper, miller, gas-fitter and coppersmith. Such diversity will never be seen again. We are still in Constable's other country, not yet the lower river with villages clogged with tourists, but upstream and in many ways a more beautiful and mixed landscape. Up on a hill ahead is Stoke-by-Nayland's prominent church tower, from which Constable sat to paint Harwich far away at the mouth of the Stour.

As the valley widens again, we slide alongside the parkland that once contained elms also painted by Constable and Tendring Hall, grand country house that burned down in the 1950s. It was home to the first Duke of Norfolk, whose house records show how he and the Duchess kept the wolf from the door a couple of centuries ago. On one particular day, when three lady visitors from London were staying, lunch was served at ten o'clock. The first course was two boiled capons, breast of mutton, piece of beef, seven chevets (seasoned meat and fish minced together), a swan, a pig, a breast of veal, two roast capons and custard; and the second course four messes of morts (salmon), six chickens, eight pigeons, three conies (rabbits), two shovellers, four sepys (oystercatchers, or seapies), a dozen quails, two pasties of venison, a tart, nuts and pears. We canoe more slowly just thinking about it. Overeating is not just a modern phenomenon.

Constable country proper is next, but first we come to the modernist pumping station at Langham's Valley Farm, where water is drawn from the Stour and sent to drier south Essex. We paddle through tightly enclosed reaches, slip under a bridge of the A12, its traffic dashing between Colchester and Ipswich, and head directly east passing through Alfred Munning's Dedham and East Bergholt where, said John Constable, "those scenes made me a painter". Suddenly the river is filled by dozens of families in rowing boats, for at two sites on these water meadows boats are for hire. It seems many of us still have a hankering to mess about on the river. Some people face the wrong way, rowing forwards, as it were. In others are packed families of three generations, arguing about being careful or whose turn it is. Hidden in the reeds and pollarded willows is a boat full of teenagers smoking furtively.

At Flatford Mill, we are allowed to pass through the working lock, and just as suddenly leave all the tourists behind and enter the empty coastal marshes. The salt of the estuary is not far away now. The hills retreat, and we drift through rough grassland, and past beds of reeds and lime-green flote grass. Hidden away on the left is East Bergholt, location of another touch of the unconventional and a special stained-glass window. Old Hall is a well-known commune whose chapel has a fine blue-and-gold depiction of St Hildegard, Abbess of Bingen, author of twelfth century Gregorian music and her philosophy of *viriditas.* One of the earliest naturalists and environmentalists. And then the fresh water runs out at a concrete barrier that marks the transition to the estuary proper. A whole river and no graffiti. This is where the water flow becomes a matter of daily tides rather than recent rainfall. Our shoulders ache and our eyes are full of light.

Leap year: I walk the Essex side of the Stour Estuary on what would have been the last day of February in three of four years. I start at the northern side of Hamford Water by the saltings owned by the local wildfowlers. At first the temperature is just above freezing, but it will rise to 14°C which feels almost hot. The hedgerows are bare of leaves, but there will be snowdrifts of flowering blackthorn, and yellow celandine, red campion, purple dead-nettle, Alexanders, chickweed and violets. I begin by thinking this day will be about filling in gaps on a route travelled by boat from Titchwell Marina to Harwich but it is more interesting than that.

Dock Lane is closed. There is still activity at the old explosives factory that now mainly disposes of illegal fireworks. I have to walk back along the main road and then drop down towards the wide expanse of saltings and islands that are the Backwaters. The morning is dead calm. Later the chorus of lapping waves will fill the air, and the rumble of heavy shipping felt deep and viscerally. I reach the sea wall, and climb up to see saltings and muds stretching away to the distant Naze tower wrapped in mist. In the foreground are Bull's Ooze, Pewitt Island and around to the south Bramble, Garnham's and Skipper's Islands. On the seaward side the air begins to ring with the clamour of gulls and waders. Then the first of the day's skylarks rises and sings to spring.

Wildfowler Albert Alcock was just 55 years old when he died, and fellow wildfowlers from Little Oakley put a bench here on the marsh in his memory. His seat is on a concrete plinth, a wreath beneath, and looks out to sea and salting, just as he would have liked. A week later, I come back on a day when a gale-force wind is rattling the windows to talk to gun-dog breeder Peter Avery about wildfowling on these marshes. We sit at the kitchen table, and Peter tells of the Little Oakley and District Wildfowlers' Association, formed in 1953 at a time when salt marshes were seen to have no formal value, being used only for wildfowl, eels and gulls' eggs. James Wentworth Day was their liaison officer, and was familiar in his chequered jacket and waistcoat. His *Coastal Adventure* drew on many of his experiences hereabouts. By the late 1980s, just a few years after Wentworth Day's death, the Association, thinking ahead in changing times, came to realise that they had no security of tenure, and decided they had to buy a piece of saltmarsh. They brought a land agent down to the sea wall at the top of the tide in a howling gale, but there was no land to be seen.

"I don't value it at anything at all," said the agent dismissively. A second valuer did his job properly, then the farmer saw an opportunity and increased the price sixfold. But the 70-member club managed to raise the money and the marsh was theirs. They have bought more since then, and now own 1,000 acres of both fresh marsh and saltings. Some have been voluntarily put into national nature reserves, and others are managed jointly with Natural England. There are also no-take zones, and always voluntary bans during very cold weather. Relations with local wildlife organisations are good, but not with the national bird one. In truth these are relatively small differences compared with the much greater threat from quad bikes and motorbikes, which have brought new opportunities to get out onto seawalls, beaches and the drier saltings.

These coastal habitats are deeply important to the wildfowlers who use them. Nothing at all can beat a night out under a full moon when the marshes are full of sky and stars. "It's not a hobby like golf," says Peter in an accent belonging to the marshes. He pats his pearly grey punt. "It's in the blood."

This land is their identity. As we stand looking at these remote saltings dotted with broken boats half full of water, Peter says the only people out here at dawn on a cold December morning would be wildfowlers. They walk and watch and observe. By Albert's bench he laughs. His son piped up at the meeting that commissioned it and suggested that the name plate should be at one end. So that there would be room for the other wildfowlers when they went too. At the beginning of the following September, Peter sends an email. He's just been out with his two sons and brother-in-law at the start of his forty-third season. They brought back a greylag and a teal. "I hope nothing will ever change," he writes.

I continue north past a sluice they call Jonty's, after Peter's wife's grandfather who lived on the landside of Strachen's Sluice. A shaft of sunshine turns the water silver and the sea beyond briefly aquamarine. More skylarks compete with one another over the inland grasslands, and then by Irlam's beach I leave the Backwaters and pass an area of concrete sea defences at Crabknowe Spit. Another managed retreat might be created here. As part of the plans to develop Barside Bay into new port facilities, the owners have to engage in foreshore recharge along this coast, and create 60 acres of saltings by breaching the sea wall. On the horizon the cranes of Felixstowe become slightly clearer and I count a parade of more than twenty. I look back, and painted along the inside of the concrete wall is a fading statement in yellow carefully edged with green, key words blacked out at a later date: *[ ] is a dirty old [ ]. She loves to [ ] young men under 20.* All these saltings and marshes, and then the first piece of concrete is plastered with graffiti. Why this message out here, and why partially censored?

Harwich from the Orwell

In the distance, the docks still appear to be in low mist. I turn inland past more wildfowlers' saltings and come to a caravan site below the wall. There is no sign of life, unless you count the repair men. *Private property*, it says. *Keep out.* Next door is the sewage works, and then I'm into the southern reaches of faded Dovercourt and Harwich. By the sports ground, and before the beach huts lining the promenade, large signs warn that we're all on CCTV. And here are more seat memorials on a patch of grass looking down on the sea. *In memory of a wonderful husband, Paul Wright, 1965, died 2004. When old times we do meet, it's then we miss you most of all.* The inscription calls to mind a grave. On either side of the seat is a pot of silk flowers, yellow and pink, and behind it a semicircle of daffodils. Empty seat, looking out to sea, missing you. The northerly one says *In loving memory of Terrence Arthur Attridge. A dear son, brother and father, 1965-2004.* Ah. Both men with the same birth year, both died aged 39. Maybe it was the same incident, at sea, where the benches face.

I sit on the second bench and look to the sea with their ghosts. A fisherman in waders is net fishing off the shore, waist deep in water. Behind me, dog owners are spilling from their cars, walking briefly, throwing a ball a few times, coming back, and driving off again. I watch for a while. There seem to be different types of owner. One does their minimal duty, and walks only 50 metres or so, throws a ball a couple of times, lets the dog scamper, and then rushes home. Another type walks fast and further, head down, let's get this over as quickly as possible too. And others are in groups of two or three, chatting, taking their time, and walking themselves as well as their dogs. There doesn't seem to be a category of dog walker who is obviously enjoying the place for what it is, but then I might be wrong. I call a halt to this amateur dog anthropology and go in search of the beach fisherman, who's laying out his net to dry.

"Caught a nice bit of sole." Soft Essex accent, eyes sparkling. Twice a day he fishes. "Mos'ly sole or flounder, but soon them bass will come."

"What else do you catch?" I ask.

"Someone caught a dogfish las' week. Thas right surprising."

"It's a lot of work for small numbers of fish," I venture.

"It is," he agrees. "It's what we do. Always have done, boy."

I walk on past rows of blue, red and purple beach huts. All possible space has been invaded by intrusive signs. There are more CCTV reminders, and then by each entry to the beach are three different signs telling dog owners to keep their animals off the beach. No cycling on the lower prom either, announces white lettering on the ground beneath my feet, though on this quiet winter morning cyclists are taking no notice whatsoever. Further up the prom are Dovercourt's two famous lights, one onshore and hexagonal, the other octagonal and 100 yards out. They're wooden structures with white slatted wood, black window frames and twisting iron staircase up the outside. The rest of the seafront is distinctive 1930s resort architecture. In a grassy sward is an oval concrete structure with long bench seats facing both inland and out to sea. On the pond I count seventy-three swans. The roller-skating rink is closed, as is a faded blue ticket shack for the Electric Palace, the oldest surviving cinema in the UK. On the high slope above this undercliff walk are modernist shapes of white stone inlaid in the grass: a boat, fish and anchor. Above is the Cliff Hotel, which has seen better days, and down here yet more signs. What to do, and what not.

At Beacon Cliff promontory an expanse of land cordoned off with rows of metal fencing separates Dovercourt from Harwich. Inside are remains of war structures – observation towers, pillboxes and operations centres. They're strange boxy shapes marked with decades of graffiti. *Nic. The Grebe. Agie. Russell was here. Adz. Ash.* It's all half hidden by drifts of burnished gorse smelling of coconut. Around the corner is a rare patch of political graffiti: *legalize the four day week*, says faded red writing. More marks and statements. We were here; we are here. And yet more reminders from the council about proper behaviour. Tit for tat, it seems to me. Of all the harbours along the east coast, only Harwich is a refuge from the scourge of a nor'easterly gale. But there has always been controversy over this harbour and the need for continued dredging to keep the channel open. The old saying "all up at Harwich" described the state of muddle and confusion that often affected the business of keeping the harbour open. But a low lighthouse was built by the sea wall and a higher light inland, both of which acted as landing lights for the harbour. Once one light was positioned above the other, vessels were assured of a safe passage into the harbour. But eventually, the shipping channel shifted, and rendered the lights obsolete. Today a deep grumbling ferry heads to the Hook of Holland, then is followed out by a clanging cargo ship. From the docks across the water comes the clash and crump of metal on metal and concrete. In the foreground, lonely fishing boats are drawn up on the tiny beach, dwarfed by the giant cranes more than a mile away.

Harwich itself is closed. Where is everyone? The main street is empty. The Chinese take-away has a St George flag in the dusty window. Around the corner is the British Flag pub. A boy sits blank-faced at an upstairs window as music plays. Two young women manoeuvre buggies along the narrow pavement. You could walk up the middle of the road without fear here. There seems nothing in the shops either. A phone shop, a closed fish and chip shop, more pubs. What would diarist and local MP Samuel Pepys have made of this? It was once an active fishing and trading town, until the fishing industry declined, and the international ferry terminal was built a couple of miles short of the town. Travellers now never come as far as Harwich proper. The railway terminus is also deserted, and I wait for the hourly train. But the guard jokes and smiles, and in twelve minutes we've passed Dovercourt's expanse of sidings and invasive buddleias, then Harwich International's characterless car and lorry parks, more cranes, and a successful crop of concrete standing. It's not the best of places to enter the country. Across the water is Shotley Peninsula, looking more desirable than this post-industrial landscape.

The train passes through Copperas Wood and at Wrabness I step off, and the sun is now shining. The short journey seems to have taken me from one weather zone to another. I look around. Not a single cloud. At the local shop the owner jumps in as soon as I open the door.

"Take that rucksack off, you'll knock everything over".

I pause, and consider walking out, or even arguing. What's the point? Welcome to Wrabness.

On the path down to the Stour, I pass curious Simpson's Farm, a block of a house surrounded by thickets of holly and bramble. Bizarrely, where there should be upstairs windows looking down on the river, there is nothing, just pebble-dashed blank walls. At the river is a field of caravans, a long-standing retreat for a group of Hackney socialists, according to Ken Worpole, and beyond them a row of wooden beach houses and chalets on stilts. It's low water so I walk along the beach. Each construction has some form of veranda or balcony, and all have evolved in different ways. A few are tiny; some vast and sprawling. Canoes, rowing boats and surfboards are drawn up underneath. There are large windows, cane chairs and enticing piles of books. It is tranquil, silent, and all the cabins are empty today.

The neighbouring cliffs are eroding, and the branches of trees down on the shore are hanging with seaweed, like washing on a line. Out by the water languishes the skeleton of a barge, and a man digging for ragworm stands and stretches his back, and bends to dig again. Wrabness was a mine depot from 1921, but was run down after the Second World War and its buildings demolished in the 1960s. The Home Office had plans for a prison, but abandoned them in the 1970s, and sold the land to a developer who wanted to build a holiday village. After rejection by planners, yet another prison plan was drawn up. Eventually in the 1990s, sense prevailed and the site became a nature reserve on what is technically a brown-field site, and now harbours nightingales, bullfinches, linnets and other local rarities. Beyond the reserve, the path becomes heavy going where the tide has come over the saltings and up to the surprisingly low sea walls. It has broken into the wheat fields, and I decide to walk in the furrows to stay dry.

A family group stands by a four-by-four down on the beach. The young father is throwing stones as far as he can. A little girl in padded jacket and Mum in red singlet are watching, and all of them are suddenly shouting, "Bella, Bella!"

A bull terrier looks up from the tall grass, growls and rushes at me.

"He won't hurt yer," yelps one.

But they grab the dog's collar and hang on tight, and I skirt by them without looking back, shoulders tense. To make matters worse, there are no more footpaths. A barbed-wire fence stretches to the shore under a stand of oak. Now, who are they at Nether Hall to stop walking along the river wall? There is no other option but to cut back under the railway line, and I walk several miles by road. Not until I reach the edge of Mistley will there be the safety of pavements. The cars and vans pass mostly with a shared wave, but a motorcycle forces me to leap onto the grass verge as it misses by inches. Then beep-beep, and the green four-by-four is surging up the hill. I wave again. Bella the familiar is staring out of the back window, unblinking.

I enter witch-finders' territory. It's homeland for the terrifying accusations of the mid-1640s which had their roots in over a century of religious disagreement. In the 1550s under Catholic Mary, 300 heretics, seventy-two of whom were from Essex, were tortured and burned at the stake. Elizabeth I gave relief to the Protestants from 1558, but sentiment quickly swung too far and many Catholics were then branded as heretics and so accused of witchcraft. In 1582, thirteen women from Tendring were put on trial and several hanged. Seven years later, another ten men and women from Elmstead and Great Oakley were accused and four hanged. Later, Charles I decided it was his duty to stamp out all forms of Puritanism, which forced many to meet privately in homes, and provoked some to emigrate to the New World. By the late 1640s, towards the end of the Civil War, the prevailing situation again allowed the self-styled and powerful to have their way. Matthew Hopkins was one, and called himself the Witchfinder General, even though there was no government sanction for this title, and he pursued ordinary people across Essex and Suffolk in a short campaign of pure terror.

It began in Manningtree. Hopkins was said to have overheard a group of women talking about meeting the Devil. At a time when the daily wage was about 2 pence, less than a fifth of a shilling, Hopkins and his two accomplices, John Stearne and Mary "Goody" Phillips, were paid up to 20 shillings by local magistrates for each witch they uncovered. Hopkins used torture to extract confessions: a swimming test (witches floated, the innocent drowned) and witch-pricking with knives and needles to search for the apparently painless Devil's marks that did not bleed. One particular method was to force victims to sit cross-legged for twenty-four hours on a table, their thumbs bound to toes with cord. In two years, he may have been responsible for the deaths of 300 people in all. Interrogations were mostly carried out at inns in Mistley and Manningtree, and formal trials at Chelmsford. Then it stopped. Hopkins' own fate is unclear. Some say he died of consumption, others that an angry crowd forced him to undergo the swimming test; in some accounts he drowned, in others he floated and was hanged. According to the Mistley church register, he was buried there in August 1647.

The odour of malt hangs on the air. Mistley, the wood where the mistletoe grew, is a former industrial town with Victorian mill architecture and the Edme Malt Extract works. Down by the river, hundreds more swans are attracted by the wastes from the maltings. It better not close, this factory, for the village emblem is a swan. The main street is Georgian and at the far end are Mistley Towers, part of a church designed by Robert Adam. Today, the churchyard and its two towers are swamped by articulated lorries parked up on the pavement. Along the Mistley-Manningtree wall runs a piece of common land along the river front known as The Strand, where geese and more swans mill for further feed from visitors. The grass is grazed bare, and scattered with feathers and white guano. An ice-cream van serves a family, others sit alone in cars eating lunch. I walk through Manningtree past new flats, a bleak industrial estate, and on to the station to catch a taxi back to Little Oakley. The station is just a few hundred metres from where we finished the canoe journey.

Easter Saturday is the earliest for eighty-five years. I've been dealt a day of snow, gales, rain, hail, and occasional sun for south Suffolk's triangular Shotley Peninsula bounded by the Stour to the south and Orwell to the north. I walk down to Holbrook Bay and south towards the Stour. Beyond the creek the sprawling naval school that dominates this north shore is asleep. Over the course of a day's 16 miles of walking, I will meet almost no one, just one couple leading their dripping dogs home in the driving rain. The tide is low, and so I walk out towards 100 or so brents, these burnt geese murmuring "brrmp, brrmp" out on the flats. I'm able to get within 20 or 30 metres before they lift, white rumps bright on this harsh morning. I stand on the mudflats and listen to the chorus of redshanks, curlew and oystercatchers, and gabbling ranks of shell ducks.

Along these shores are yet more eroding cliffs, the long roots of blackthorn hanging down and sprouting. I head east toward the sea, past saltings at Johnny All Alone Creek, which then spread out into the river at Erwarton Bay. The wind is from the north, and I'm partly protected first by cliffs and then by fringing oaks on the tops. I walk along the wet beach, and then suddenly find myself caught in saltings. Griff Rhys Jones writes that you can run across the muds, if you move swiftly. But it takes some nerve, for it would be a lonely horror to be stuck fast. I could go back, but decide there must be a way through the spiky cordgrasses and slushy muds. I have no marshman's splatches or pattens, so increase speed to stay above the crow black mud. Suddenly I find I have to run fast: slap, slap, slow black, slap, slap again and

Pin Mill on the Orwell

there ahead are the oaks at the cliff base where the sand should be firm. I leap at the end, and straighten to catch my breath, and decide to go up after all. The wheat fields are a saturated bright green, no signs left of the frost damage that scorched them all yellow a few weeks ago over on the south shore. A hailstorm races across the open fields from the church at Erwarton, where they say Anne Boleyn's heart is buried, and batters into me here on the windward side of the trees. I struggle past a farm with meagre cottages of grey concrete walls and metal-rimmed windows, a green scum of the *Pleurococcus* algae covering old cars and abandoned caravans. It's cold comfort on this bitter day, even though on the riverside there must be glorious views over the water.

This is land where they make sailors. I have come to Shotley Gate itself, up past woods and streets of houses with no views, down to the piers where all is desolate and abandoned. The sky is heavy and rain lashes down. One pier is blocked off by fencing, and inside is the only graffiti of the day. Around the corner is a surprise: investments have upgraded the marina road and promenade, and shiny metal bars separate the walkway and churning harbour water. Here are the hurrying dog walkers, smiling briefly, pushed on home by the wind. Ahead are hundreds of yachts, and up on the hill a couple of Martello towers and the 150-foot mast of *HMS Ganges* naval school. The Black Prince was left standing here by his father Edward III in 1340 when the King set off for the Battle of Sluys. Docked across from Blood Point is a vast ship, China Shipping, with hundreds of containers perched on deck. Out in the harbour, the ferry to the Hook moves east, silently downwind today. Seasickness seems likely. I turn the corner into the Orwell to face north-west, and instantly am leaning into the teeth of a gale. Opposite, the port suddenly terminates at Fagbury Point, and beyond lie marsh and fields, and that's the end of the industrial shore.

Waves are crashing onto this shore. White horses dance down the Orwell, and the water is muddy brown. Now a horizontal hailstorm joins the attack and I have to hold up both hands to protect my face. I huddle behind the sea wall, and watch a flock of greylags that have all turned head first into the wind. I continue along Shotley Marsh and then towards Crane's Hill saltings. The squalls whip by, now briefly bringing blue sky and sun, and again come the boiling, racing clouds shedding more heavy rain and pelting hail. Then the reward: I am walking with inned marsh to the left and yellow-green saltings to the right, when two slender chestnut weasels scamper down the sea-wall path towards me. They stop abruptly, and together draw up on hind legs. And I'm thinking of the story I read just last night by Annie Dillard. *Living Like Weasels.* The unknowable eye of the wild animal. And here are these weasels and their white bellies on the wall and me. I bend down slowly and we look at each other. And that is all there is. We are frozen for a moment, face to face, absorbing, and then the gang is gone, bounding down the inside of the wall, the long grass rustling at their passing. Dillard says of her weasel, "I don't remember what shattered the enchantment."

The footpath slips through a garden, and then climbs into Cliff Plantation's carpet of green bluebells that will flower soon and turn this wood smoky blue. The path drops down towards Pin Mill, and I am anticipating a halt at the riverside Butt and Oyster, formerly favoured by smugglers and cargo-runners. But first there is a line of thirty or so eccentric barges and floating homes, gangplanks cobbled together, plants in pots, bicycles and old motorcycles at the ready, *Pleurococcus* everywhere. In this wind the tenders are being battered by waves, rising, falling, crashing against the solid barges. At the end, there is no way around the point to the pub door, unless by swimming. I climb the 30 metres to the top, creep under a fallen ash, and then walk back down. The bar is filled with acrid woodsmoke from a blazing fire, and from the windows is a view of the foaming river. A century ago, William Dutt sat here and watched "brown-sailed barges slowly tacking up and down the river". Today, nothing moves on the river. In front of me, I have a pint of Best, and my face tingles and eyes smart.

Somewhere ahead is the great road bridge. I pass a school on the hill and white modernist Harwich Yacht Club. Like on the Stour's southern shore here's another batch of sealed-off coast and I have to press inland again. There are heavy clouds over open fields towards Freeston, but the four-storyed hexagonal Tudor folly is hidden down in the woods. I walk up an enclosed path where a photograph is pinned to the fence: Jack the donkey lived here, it says, but has passed away. At the Boot Inn, there is a sign, and I realise a bus could take me back to Holbrook if I'm lucky. I walk down the hill and the land opens up, and there is the Orwell Bridge carrying the A12 traffic around the south of Ipswich. Says the publicity, it's the largest prestressed concrete span on a bridge in Britain. We've driven over this bridge so many times, but never passed beneath. I stand directly underneath, looking across the saltings, now disappearing under the rising tide. A single red-brick house is dwarfed by the grey buttresses and oppressed by the rattle of traffic far above. It must be a kind of hell to live down here. And then, on time, comes a red bus. I put out my hand, and it stops. I sit and watch the blurred view through the condensation. Time warps again. Walking takes all day, journeys by road eat up the earth. The next morning, it snows heavily, silencing all the land.

N
W
E
S
SUFFOLK
ESSEX
RIVER CROUCH
Head
FOULNESS
ISLAND
The Broomway
MAPLIN SANDS

Chapter 8

ALDEBURGH
ALDEBURGH BAY
Slaughden
RIVER ALDE
ORFORD
Orford Ness
Pagodas
SHINGLE STREET
Rendlesham Forest
BAWDSEY
HOLLESLEY BAY
NORTH SEA

Orford lighthouse

Chapter 8
Strongholds

The quay is quiet. Too quiet. Over the water a herring gull screams. It's early on a summer's morning, and we're waiting for a boat. Over there is Orford Ness, the largest vegetated shingle spit in Europe, long held by the military who found the remoteness suited them well. Orford was a port once, but soon after the medieval castle was built the shingle started migrating, and pulled the Alde River mouth 10 miles south. On the land side, Orford now hides behind the great Tunstall and Rendlesham Forests. A boat of night fishermen appears, ties up, and the men fill their car and depart. Still we wait, and then call, and I'm told that the warden due to give us the lift no longer works for the National Trust. My heart sinks, and I wonder how yet another ferry ride has failed to materialise. But no. Another agrees to run us over, and a little later we are on a boat being drawn into new excitement. Last night, a routine moth-trapping exercise out on the marshes captured a death's-head hawkmoth. The largest moth in Britain, and a rarity. We wait for a boy from the village, who arrives hopping in anticipation, and soon after crossing we are at the offices peering into a translucent plastic box that has been in the fridge overnight. Inside is the moth, dusty yellow and black, with skull and scissors at the top of the thorax. It's named after Atropos, one of the three Fates, who chose the means by which mortals would die and used her shears to cut their thread of life. And this is why we're here; to see a land devoted to ways of killing.

This east coast has always been in the front line of national defence. During the ice ages, it was linked to Continental Europe by land bridge, and this was how the first modern humans came across as they pressed north, eventually to displace the long-resident Neanderthals. Much later came Romans, Angles, Saxons, Danes-Vikings, and finally the Norman armies. After 1066, though, no further invasion efforts were to succeed. Many tried: the Spanish Armada in 1588, the Dutch skirmishes of the 1600s, Napoleon's expansionist plans of the late 1700s and early 1800s, the repelled efforts of Nazi Germany in the 1940s, and finally the cold war threat of the former USSR. As a result, the coast is littered with forts, castles, pillboxes, mine towers, beach defences, tank traps, radar stations, masts, weapons establishments, testing facilities and many other mysterious buildings. They help to define the character of this coast.

We leave the moth men, and set off along narrow paths through the shingle towards the sea and its red-and-white lighthouse and neighbouring black beacon built in the 1920s for an experimental navigation system. Orford Ness was long-forbidden territory for any but the military and wildlife. It is not so famed as Derek Jarman's Dungeness or Dorset's Chesil Beach, but still it's 16 kilometres long and covered by sea beet, sea kale, orache and yellow horned-poppy, thousands of ground-nesting birds, and all kinds of unexploded ordnance. Hence the good advice about staying on the paths. The lack of visitor disturbance also helps the little terns as they are easily scared off their nests. The shingle ridges and valleys have been deposited by centuries of sea action, and each ridge is an ancient shoreline. Some are rare acid shingle with communities of sea campion and stonecrop. There are saline-percolation lagoons, too, with cockles, endemic crustacea and the rare starlet sea anemone. Seaside gulls abound, as we found earlier, but curiously they are a relatively new phenomenon. Before the late 1960s, no black-backed gulls had been recorded at Orford, and the hefty herring gulls only appeared in the mid-1990s. But these gulls have three main feed sources: shipping that chops up fish by propeller action, refuse tips, and free-range pig farms, and now there are many thousands.

The War Department acquired Orford Ness in 1913, and built one of the first airfields in the country. It saw the world's first parachute jump, only a few hundred years after Michelangelo's original parachute design, and was the location during the First World War of the only attack by Italy on the UK: a single biplane being shot down on the beach. For decades, large parts of the spit were used as firing and bombing ranges. The first radar experiments were here too, in the mid-1930s, before Robert Watson-Watt moved his team to mainland Bawdsey Manor. In the war, Barnes Wallis used nearby Shingle Street to test some of his bouncing bombs. At the height of the cold war, there were 600 people stationed out here. At the north end of the spit was Cobra Mist, an array of 200 masts over an aluminium net that was an attempt to develop over-the-horizon radar. It never worked or, as some have speculated, was just jammed by the Soviets. It closed in 1973, and just eight masts now remain for use by the BBC World Service. The Ness is most infamous, though, for atomic weapons testing. In the 1950s William Perry led the environmental

Orford pagodas

testing of nuclear bombs in six underground laboratories, now called the pagodas for their mix of functional modern and oriental design. One of these was the dewdrop tower, where conditions were so cold that scientists wore special-issue duffle coats. The rum ration for RAF staff was last awarded on Christmas Island (where it was very wet) and on Orford Ness (where it was so cold). The pagoda facilities also tested rocket-propelled weapons and freefall bombs, and other parts of the spit were used as a bombing range well into the 1970s. Today, the craters remain and rusting ordnance is scattered across the shingle.

Earlier in that wet summer, I went out walking under this atomic sky with Rob Macfarlane and Leo Mellor. We meet at Ipswich station and come to the Ness for a tour around the pagodas, Johnny Cash ringing in our ears. Grant Lohoar is the teller of stories, and reminds us that lies and myths are part of the territory, and the difference between the deliberate and accidental are often blurred. No one can be quite sure of any truths, pieced together as they are from personal testimonies and the few remaining written records, which may have been mistruths anyway. The Information building has a pocked sign prohibiting both photography and sketching. The pagodas are piled high with shingle on the outside, but close up you can see that any rumours about deep tunnels and networks of bunkers are plain wrong, as they'd be far below sea level. The iron gates are rusty, walls an institutional green, peeling, flecked and dusty. It reminds me of the abandoned buildings in the town of Pripyat, close to the Chernobyl power plant complex. Intestines of wires and cables protrude from ducts, and a bone-deep cold grips the air.

Grant peels back a tarpaulin and shows us a gleaming white bomb on an olive trolley. It's the WE177A, 24 feet long by 7 wide, weighing 4½ tons, and eight times more powerful than the Hiroshima bomb dropped by Enola Gay. It's in chillingly perfect condition. This model was first tested on August bank holiday in 1956. My father's good friend from the RAF, Pete Wilkes, hero and uncle to us boys, was a flight engineer sent on aerial-sampling missions through that mushroom cloud. He died of a sudden-onset brain tumour in the 1980s. We walk around this secret pagoda room, and Leo notices an official stamp and letters in black on the glossy white bomb. There's a British Standards paint-specification number. "No," we say, laughing. You can drop the nuclear bomb, but the paint will be safe. Marvellous. Those weapons developers also loved dreaming up fanciful nicknames. There were dozens: Yellow Sun and Red Beard tactical nuclear bombs, Blue Peacock and Brown Bunny nuclear landmines, and bombs named Blue Danube, Blue Rosette, Green Bamboo, Orange Herald, Purple Passion and Red Snow. The more descriptive Cudgel and Fishfryer didn't get past the proposal stage.

The most famous myth for this region was a supposed UFO landing. The Rendlesham Forest incident occurred on Boxing Day in 1980, seven years before the pine forest would be flattened by that unforecasted October hurricane. Residents of Sudbourne village reported seeing lights in the trees and a strange shape in the sky. Not unusual given all the military airfields near by. Servicemen went into the forest, and saw more lights, one from a single source pulsing blue and red in a yellow mist. The Suffolk police chose to call it a UFO; the deputy commander of the base, Colonel Charles Holt, wrote a memo admitting the presence of unexplained lights, but hinted at no explanation. Some say it had to be Orford lighthouse reflecting off low cloud and mist. Others like to believe in an advance guard of aliens. But who knows what was happening out here on the Ness that day after Christmas? No one's admitting anything.

On the summer morning of the moth, we walk north of the pagodas, across the shingle ridges, past purple sea pea, once a bitter famine food for people of this coast. Some shingle furrows, strangely bare of vegetation, are full of tangled remains of wire, shells, and strips of rusting metal. We find ourselves looking for the hares, but don't see any. Like the foxes, they swim across at low tide, or make their own way down the narrow spit from the north. This is also, oddly, a place where urban foxes are often released. They are trapped by councils and welfare groups who think they are doing the foxes a favour, but sadly the foxes can't compete with the canny locals. The wardens say it would be kinder to kill them rather than go to all that effort.

After the hooped lighthouse by the shore, one of the oldest lights of the whole coast dating from 1792, we turn north and disturb a seal pup. It scuttles down the shingle, splashes into the water, turns to watch us, tipping its head on one side. It's 6 miles of shingle to the lost village of Slaughden at Aldeburgh. Inland the marsh music of curlews and oystercatchers echoes over the site of Cobra Mist where only US personnel were permitted in the inner sanctum. This is hard walking on the large stones, feet slipping and pushing down to ankle-depth, but on we struggle. Before the town, weighed-down beach fishermen are tramping the other way, and then we arrive at the only clover-leaf-shaped Martello tower on the coast. Most of the forty-five towers are thick-walled semi-conical structures of two floors. They were built in the early 1800s, modelled on a defensive structure at Mortella in Corsica, the myrtle tower, which had repelled two British Navy ships. Strangely many were not completed until twenty years after the threat of invasion by Napoleon had evaporated. They're survivors.

The military might have deserted Orford Ness but they retain control of a 6,000 acre island some 35 miles south-east of Orford as the gull flies. Back in south Essex, location of an overnight stay early in these walks, Foulness and its neighbouring islands of Havengore, Potton, Rushley and New England were also bought by the War Department in the First World War. Sea walls had been raised in the late 1200s for marsh grazing, housing appeared by the 1300s, and the population reached 900 in the late 1800s. Today, the island and all properties, save for the church and rectory, remains in the hands of the now Ministry of Defence. It is used for all kinds of secret stuff: weapons development and testing, the destruction of old ammunition, and more atomic testing. Explosions remind residents up and down the coast that there is still an active military out on this distant patch of the county.

There was no bridge to Foulness until the 1920s. Before then the main road to the island was out on Maplin Sands and sluiced by the tides twice a day. It's called the Broomway, and is quite simply one of the most remarkable roads to any island in Britain. One early August morning before the long walk, I join a group guided by Brian Dawson and local residents Anne Chittock and John Burroughs to walk the Broomway. We start from Wakering Stairs, near the great brickworks that were once fuelled by London's rubbish, not far from a Celtic burial place. To the south, the vast Isle of Grain chimney looms. We stand on the sea wall looking east to the sun and sea. How precisely, everyone is thinking, will we walk on water? The route of the Broomway starts at the concrete hard that has been laid over hazel wattles, and heads disconcertingly directly out to sea. The tide is out but shimmering puddles on the sands make it look as if we are going to advance into deep water.

We walk half a kilometre out from the sea wall, then turn north-north-east to proceed parallel to the coast. For centuries hundreds of brooms were planted at roughly 30-yard intervals to mark the safe passage across these sands. Some say these were actually upside-down brooms for sweeping floors; others that they were switches of broom plants. Either way they were called mapples. It's not clear which came first: were the Maplin Sands named after the mapples, or the mapples after the Sands? In 1904, Herbert Tompkins on his *Marsh Country Rambles* recommended that "it is best to keep near the brooms, which, as the tide ebbs, gradually lift their heads and appear as a line of dots, some six miles long". At the time, the postman was still driving his pony and trap across the sands. Although Tompkins said that Foulness itself "affords a dreary prospect as you approach it from the sea", looking away he saw that "a thousand tiny pools glittered in the morning sun".

The Broomway dates from at least the 1400s, and access is from various headlands – Shelford, Asplins, Rugwood, Eastwick and finally Fishermen's Head a couple of miles short of the northern tip of Foulness. As this was the main road onto the island, local people needed exceptional knowledge of tides and the sands. Nonetheless, many did drown out on the Broomway, and it was known as the most dangerous road in England. Bob Crump of the Foulness Conservation and Archaeological Society records an 1899 letter from the Reverend R. H. Marsh, who travelled this route in early October from Wakering Stairs in pouring rain, apparently laden with books and walking barefoot. The sands were dark, and he had to navigate from broom to broom as a thick mist descended. As he arrived, "the island was so dark, I could see nothing". He wrote that he then saw strange green and yellow lights glowing in the mist, echoes of other mysterious coastal lights, and of invaders and wreckers too.

It is difficult today to imagine the route in dark, wind or rain. The distant Shivering

Wakering Stairs to The Broomway

Sands are rightly named, as the morning light dances on the water, and container ships levitate in heat mirages. I remove my shoes like the good Reverend, and feel the hard sand under its film of soft mud. I see fingerlings of perhaps goby or wrasse in the warming puddles, and wonder at their capacity to cope with huge daily temperature shifts. Near to the shore are the eelgrass-covered sands called the blackgrounds. This is the largest stand of dwarf eelgrass in Europe, and the *zos* is actually dark green. We're walking on the coastal sands called the Swatch. We pass the skeletal remains of a 60-foot cruiser and then a small trawler locked in the sand thirty-five years ago, both of which grounded on their last sea-going day, and never floated off. Their rib timbers are reminders of the treachery that lies under these waters. John, one of the island's arable farmers, tells me wistfully of coming out onto these sands under a full moon in late autumn to hunt wigeon. He brings a board to use as a shooting stick, and leaning against it feels he could be on the far side of the moon. Magical, he says. Standing here, it is easy to see why there was such opposition in the late 1960s to these sands being a possible site for London's third airport. Luckily, Stansted was chosen. Perhaps Foulness was saved from the planes and industrial development by the birds. The name derives from "fulge-ness",

Saxon for a promontory of wild birds. I mention the airport plans. "Wash your mouth out," says Anne. "We don't talk about it." It took years of effort to win that fight. Out on the swatch today a herring gull, sick from salmonella or botulism picked up from the tips at Mucking or Pitsea, is being mobbed by other gulls. Its squealing echoes fall across this vast sandscape.

After two hours of hard sand and warm puddles, we come onto the island at Asplins, past a stand of golden samphire full of sawing grasshoppers. Here are red danger signs, great 16-inch shell cases and a line of truck-sized cages used for destroying ammunition. The charm of this coast is that one minute all is salt and sea air, the next you are in warm dusty fields of wheat ready for harvest. A marsh harrier rises slowly out of the corn, a looming dark shadow, and yet we can still hear the piercing calls of waders from out on the sands. In a sweet hay barn, we investigate kestrel pellets under a box. A cockerel cackles, and swifts chitter on the breeze. A mosquito settles on my arm, then another. There is a distant crump of military activity. Nature and war, side by side. We walk up through Churchend village and its wooden houses covered with rambling red roses. Only occasionally does a car pass, and I find myself surprised they are not vintage models. The pub's walled garden is rich with every vegetable, another reminder of the island culture of self-sufficiency. In the 1800s, bare-fist-fighting competitions were held here by the pub. John Bennewith was a famous local champion, and fought the Infant, Giant, and Bullock's Bones. Today the landlord is seriously grumpy with visitors: don't do pineapple juice, he says; don't do soda. But as I sit on the wooden bench he calls across to the locals, "He's doing his notes already," nodding at me with a broad smile.

We cross more shimmering fields, past goldfinches squabbling over teasels, and arrive at the dock on the western side of the island. We intend to cross the Roach to Wallasea, though now there is a hitch as the boat fails to arrive. Those ferries. Brian's colleague John comes to the rescue and starts to take everyone across in a small boat. I cross with Brian's daughter, Teri, and we walk down past the 100-plus hectares of new wetland created by five breaches of the sea wall a year earlier. Originally there were some 35,000 hectares of salt marsh in Essex; now only 2,000 remain. These new wetlands are an offset for Keppel Bank cargo terminal in Kent. But we see from a distance that many of the other walkers are still stranded: two outboard motors have failed and eventually John's trusty lifeboat is pressed into action. We spend a couple of hours in the late afternoon sitting on the dusty grass, watching the tide rise, gradually filling the new Wallasea flats with bright-blue sky. You get a lot of sky in this water.

When Ken Worpole and I stayed on Foulness after the colourful chaos of the seafront, we had to obtain permission from Qinetiq, the company that now runs Foulness for the MOD. Local people say that when Foulness was run by the old regiment at Shoebury they could call up the Colonel-in-Chief if there were problems. Farmers especially would prefer a bit more discussion on the use of live shells when their combustible cereal fields are ready for harvest. Fires are common. Today, they have to interact with a faceless company whose helpline is answered by machine rather than real people. Everyone has to sign in and out each time they leave or arrive on the island, and residents have to alert the authorities when they have visitors or deliveries. But we get every help from Nicky the community liaison officer, and are able to stay overnight in a tiny room at the pub.

That evening four islanders, Sam and Mary Self, and Roy and Linda Ducker, take us back with stories of the 1953 floods and the war. Their accents are of the marshes, coastal Essex rather than out of London. Once a Heinkel came down on the sands at low tide, luckily for the crew, and Sam remembers the pilot being driven off clutching his knee, Luftwaffe wings glinting on his chest. Spitfires regularly used the wreck for target practice, and even though they were forbidden by parents, all the local boys played in the remains of the plane, running for their lives each time they heard the distinctive Rolls-Royce engines roiling across the sands.

The Second World War ended over sixty years ago, but remains a central memory for many people of the coast. It defined early lifeways and formed identity for many. The painful evacuation of children and the splitting of families, the constant fear of night-time air raids, the rationing of foods and daily search for anything edible from the wild, the blurring of fact and fiction by authorities. From 1939, blackout precautions allowed personal homes to be searched at any time and no one was permitted to keep more than one week's food supply at home. There was no fishing or fowling for boat owners. If their vessels were not hauled out of the water, then they were sunk. There was constant movement of people too, and those who could remain in one place were lucky. Thousands of airmen from many countries came to East Anglia's 109 airbases, and later Italian and German prisoners of war were housed in holiday camps and in rural communities. The German POWs on Foulness, said Sam, worked the farms. Each had a large spot on the back of his shirt, and most were entirely content in the latter years to stay right where they were, seeing out the war growing food rather than risking being posted to the Russian front.

London's first evacuations occurred as the war started. The government was worried that bombing raids would cause mass panic. No one had yet conceived of the Blitz spirit. Plans were made to move huge numbers of children. "No such undertaking," said Chamberlain on 3 September, and the evacuations began. On the first day, 14,000 were moved by train and paddle steamer from London to Colchester, Felixstowe, Lowestoft and Yarmouth. Each subsequent day, another 4,000 were evacuated. Families were broken up, and remained separated for years. In early 1940, it was realised that the coast was even more risky than the cities. In spring and early summer, a third of a million children and adults left the east coast for the Midlands and Wales. Children carried labels around their necks and were allowed one small piece of luggage, making seemingly endless journeys on wheezing, clanking steam trains, ever further from home and the comfort of family. Then at the end of the line, they were parcelled out to local families. My own mother was luckier, following her mother to Scotland, although she still attended fifteen different schools by the age of 11. My grandmother was WRENS service manager of an officers' mess, as she had the experience of working in a pub, and loved the war.

There were daily migrations too, particularly after June 1941 when the first doodlebugs, or V1 (Vergeltungswaffe) rockets, were launched. These were 2 tons in weight, 25-feet long with a 16-foot wingspan, and a range of 150 miles, later extended to 250. Up to 100 each day would come droning over the coast, and though they didn't kill many people, they did cause enormous anxiety owing to the random nature of their eventual fall. As the drone cut out just before the final descent, it would leave everyone wondering about its target. One V1 sliced off a corner of the roof of the Still and Star in Aldgate. It became common for city people to leave the urban areas early in the evening to escape into the country or woods for some sleep. Epping Forest had a whole tented community of escapees from the rockets and their dismal droning. When the V2s, six times heavier and with a much greater payload, were first launched, government kept silent to maintain morale for as long as possible. More than 1,000 were to land on England, 450 on Essex alone.

Meanwhile, the whole seashore became a vast defensive stronghold. Every beach had iron spikes anchored in concrete and pointing to sea with concrete blocks set between. Inland were roadblocks, pillboxes, anti-tank ditches, anti-aircraft batteries along both coast and estuaries, and thousands of miles of barbed wire. A giant boom was built across the Thames from Shoebury to Kent. And all these waterfront facilities were staffed by nervous and expectant mixtures of regular troops and the Home Guard. Everywhere, too, there was deception. Factories were camouflaged: the roof of the Ford plant by the Thames at Dagenham was painted with a marshland scene. There were decoy sites too – decoy bomb dumps, decoy submarines, and decoy aircraft, all designed to lure enemy bombers away from towns and other targets. Decoy towns called Q-sites were constructed in rural areas. There was one on the marshes at East Mersea, and another between Nacton and Levington, and a further one at Walton to distract bombers away from the explosives factory at Bramble Island. When planes were heard, soldiers rushed out and lit up the area to attract the planes like so many hawkmoths. The most ambitious decoy effort was the creation in 1944 of a 50-division shadow invasion force in East Anglia, with false boats only half hidden in the creeks and rivers from the Yare to the Orwell and Deben. They were crewed by overage or wounded soldiers, and German bombers were permitted occasionally to intrude, leading them to think this was the D-day landing force in preparation, thus distracting them from the real one far away on the south-west coast.

Shingle Street on this south Suffolk coast is the site of an enduring invasion myth. The experimental radar team had moved from Orford to Bawdsey Manor, their 250-foot towers able to track aircraft 75-miles away. They must have had a pretty good idea of what was happening on the coast. Yet nearby motorists often complained of engine failure, and alarmingly pilots made similar complaints about their engines occasionally cutting out. There were, too, tales about fishermen's exposed skin becoming blistered. There was talk of secret rays coming from the radar station. Guarding the remote marshes and beaches was a fearful task, as invaders were expected at any time. In December 1939, a patrol at Shingle Street just disappeared, and was assumed taken or killed by an enemy submarine. The famous Shingle Street incident occurred five years later when bodies were washed up to tangle amongst the barbed-wire and steel and concrete sea defences. Some say they were German soldiers intent on invasion, and burned by experimental flame guns. Others say that they were British seamen from a mined destroyer. Or were they victims of an experimental chemical weapon? The official papers on the incident were supposed to be kept secret until 2014, but were opened in 1992 after much agitation. And they revealed? Nothing at all. But it will not be long before no one living on this coast will have personal memories of that war, and its deep shaping of values and places. The blurring of truth and myth will continue.

N
W
E
S
SUFFOLK
ESSEX
HARWICH

Chapter 9

Snape Maltings
HAM CREEK
ALDEBURGH
ALDEBURGH BAY
IKEN
Slaughden
Staverton Thicks
ORFORD
Orford Ness
BUTLEY CREEK
Boyton Marshes
RIVER ORE
Hollesley Bay Colony
SHINGLE STREET
HOLLESLEY BAY
BAWDSEY
RIVER DEBEN
NORTH SEA

Snape maltings

Chapter 9
Shingle Shore

The ferry smacks headlong into the steep bank of shingle with a crunch. I leap out and immediately sink. I press up the slope, feet slipping back into the stones, and behind me the boat is roaring away into the blue harbour. I had waited at Ha'penny Pier across in Harwich, buffeted by a rising nor'westerly, watching a yachting family prepare for their day. Then the ferryman arrived on time. Just the two of us embarked at the early hour, and set off to find Suffolk. He was once a fisherman. We chatted about this cold and wet summer. Not global warming, he objected, just the weather. He fished at a time when there were twenty-six boats out of Harwich. Now only five or six remain. The old enemy is to blame: the European Union. It's true that fishing communities have suffered at the hands of the Common Fisheries Policy with its disregard for social side-effects. But then again, if there are no fish, they can't be caught. The ferry slapped across the waves, nosing around the vast cliff-faces of container ships heading for the docks. Yachts zipped over the waves in the bright sunlight. Two million containers brought here each year carry every kind of consumer good from all over the world, yet many of the departing containers are full of rubbish, or even empty. Not much to be proud of.

A taxi takes me the 5 miles along Felixstowe's front to the next ferry at the Deben. Technically it's a cheat, but I have come to accept that it's impossible to cover every inch of this coast. A 1950 guide to the Suffolk coast describes Felixstowe as an "air-swept place, sunny, fresh and open, pure and sweet". I walk down to the fishermen's shacks by the ferry. I remember this place. Some thirty years ago, I used to come here with friends, and we'd lounge about in the glassed-in veranda of a wooden chalet, listening to music, or spend hours testing the quality of Southwold's finest in the pub. Sitting in the sunshine outside the café today, I wonder what happened to them all. We think at the time we'll stay in touch forever. A fisherman wearing a peaked blue cap steps out of a hut and surveys the swirling waters and shifting sandbars. Three generations of a family walk to the ferry sign, and wait in line. Is this is a ferry that works, or just pretends to?

An open boat motors into view exactly on time, this ferryman in smock top, puffing on a pipe and staring to the far horizon, and looking like an archetype in a P. H. Emerson photograph. I grab my rucksack, and a few minutes later we are across the Deben, and that's the third estuary of the morning crossed. The family heads inland on their mission, while I turn sharp right, and head for the waves singing on the shingle, stepping around clumps of yellow horned-poppies. This, I soon realise, is going to be tough. To port, slipping slowly backwards, lies fenced-off and abandoned Bawdsey Manor, and now a forgotten backwater since the MOD moved out in 1994. Holes have been punched in the fence, and the charcoal remains of barbecues can be seen in the sand between the clumps of tamarisk. I begin to search for the stones that are most supportive, but it takes me more than twice the time to cover a mile than on any other surface on the whole coast. I learn the subtleties of toffee-coloured shingle species, and then find a grey-blue line of larger stones that seem to take my weight well.

The shingle goes on. Swinburne didn't care for this stretch: "Miles and miles, and miles of desolation! Leagues on leagues on leagues without a change!" That's unfair, yet it still seems to take an age until the footpath diverts inland. There's coastal erosion ahead, say both the map and the signs. I take the path for a few minutes, then pause, look up at the sky, and turn back to the beach. I don't feel like being defeated by some crumbling cliffs. Thirty metres of these fields were washed away last winter. Unlike Essex, there is little hard sea defence on the Suffolk coast. This coast needs help, but for now there's no more money. Heavy granite boulders from Norway have been cast around Bawdsey Ness, and now agencies say that is it. Local farmers have identified some land that could be sold off for housing, and this could pay for the sea defences necessary to protect their farms. But something doesn't seem quite right. I scramble over the slick blue-grey rocks and their sharp ridges, and climb to the grass platform above, where moving north I skirt around a Martello tower close to the cliff edge. I hang onto fences that run out into thin air, swinging around them, and then tramp into forbidden zones marked with bright-red warning signs. But I survive and come upon a great abandoned expanse of concrete gun emplacements and grey lagoons. The sea is largely deserted too, save for occasional container ships. For centuries, this coast was a major trade route, coal coming from the north-east for heat and gas manufacture. In 1850, 10,000 seamen were said to be employed bringing 2½ million tonnes of coal to London alone, ships making

Shingle Street

8,000 trips up the Thames each year. An observer at Southwold once counted the sails of 400 colliers passing on a single day. The North Sea was a busy motorway network; now all the trade has migrated inland.

The low cliffs abruptly give way to shingle banks and inland marshes. I can see that there's a path on the inner sea wall all the way up to Shingle Street. Ahead of me, a kestrel appears. It hovers, watching the grass intently for signs of prey, and then we seem to connect and move in tandem. As I come closer, this rufous bird swoops to one side and then snaps into position a little further ahead. I move, and it moves. We dance our way up the path for about a mile, but it catches no lizard, shrew or vole. It just hovers, watches, and moves, and I just keep on walking, walking, walking. Then without warning, it flips away with a "kee-kee-kee" and blurs across the fields. "Come back," I whisper. I am all alone to walk the final stretch to the beach houses and bungalows that make up this tiny settlement. Each is different, organically divergent over time, and all were smashed and filled with shingle that night of the crashing waves of 1953.

I am standing by a tamarisk bush intertwined with a pine, gathering my thoughts, when out of the foliage flies a peregrine, and it flees up and over the lavender bungalow, completes a circuit with a few stiff wing beats, and is away too. The fastest animal in the world. I take a few moments to catch my breath. I can hear the cocktail crush of stones under the distant waves. Here Sebald bleakly thought "the silence was profound", and imagined himself "amidst the remains of our own civilisation after its extinction in some future catastrophe". Solitary walking can lead us to find emptiness within and without. I call Brian Johnson, another ferryman, and we meet to arrange a time to cross the Butley River by the occasional ferry. It is one o'clock, and I reckon on doing the 5 miles of intervening marshes and saltings in a couple of hours. He'll drive there in a few minutes.

First I walk to the terrace of white coastguards' cottages to search for the shell sculpture. Brian thinks it was created by local people, a job for some children: collect up all the whelks you can find. Later he discovers that Lida Kindersley, who visits the Street regularly, was the originator of this land art. She began with a line of bleached shells around a favourite plant, and then extended it. Others have since added more, and remarkably it survives and appears not to have been vandalised. I gaze along the 15,000 shells stretching several hundred metres, then walk towards the sea to find that there have been adaptations since I was here with Rob and Leo three months before. There is now a devil figure with horns holding the line and a sun off to one side, its beams reaching to all points of the compass. The sculpture is evolving here on the beach. There is no signboard: we must fill in the story ourselves.

I dwell too long. I'll have to step out to make the deadline. This has happened before. I make an appointment to cross a river or estuary, and then worry invades a period of walking. At the mouth of the Ore, the landscape is suddenly wild. The water is rushing inwards as the tide floods, and all around are low-lying saltings, now bright against brooding black clouds. I walk fast up the inside of the Ore, past the site where Julian Tennyson nearly drowned in the 1930s when his boat capsized, now Simpson's Saltings nature reserve where there are rare lichens amongst the cordgrass. Then I pause to note the path up the hill towards Hollesley Bay Colony, formerly another farm which trained men to become good colonial agriculturalists. Established in 1887 as a showpiece in orchard and soft fruits, it was bought in 1938 by the prison service to house young offenders, and then developed as a prison. Today, it is notable for its herd of one of the rarest animals in the world, fewer in number than mountain gorillas.

The first stud of Suffolk Punches was established here, and these horses became an icon for the county. Suffolks are tall, chesnut in colour, as they are locally-called, often with a white star or blaze on the face, and have long been admired for their docile temperament. They can all be traced back to a stallion foaled in 1768 called Crisp's Horse of Ufford. These giant horses were bred to work the heavy Suffolk clays, but went into almost terminal decline as tractors and combines came after the war. By the early 1960s only five breeding farms remained in the county, one of which was Hollesley Bay. Luckily, the prison service had long recognised the therapeutic value of horses in resettlement programmes, and somehow the colony escaped the clutches of the accountants up in Whitehall. But in the 1990s, the bean counters prevailed, and the prison service decided to sell the farm and dispose of the animals, believing now it was demeaning to work the land. More importantly, the land would raise real cash. Suffolks were suddenly under serious threat.

I wait for spring to meet John Marsh, one of the lieutenants in a latterday horse rescue. In 2002, the Suffolk Punch Trust was formed as a charity to save the colony, contribute to prisoner rehabilitation, and develop nature-based education for children. They struggled to raise the money, saw the land price increase, were turned down by the Lottery Fund, but eventually did get control of the farm, horses, and dilapidated buildings. We walk around displays of fine carts and farm equipment, a room of magnificent brasses and harnesses and then are about to set out to walk the farm when John is told that a sum of petty cash is missing. In the last six months of their sentence, prisoners can choose to work on the farm as part of community service. The Trust takes two to eight a day; some are good with animals and land management, says John, many quite useless. The best are gypsies and travellers, who still today know their horses. For the rest of the shortened walk, he's distracted. It might be stolen, or maybe just misplaced. He needs to find out.

We have time to beat the bounds of the nearer fields. I've never really taken to horses, and they seem to know. This is strange because Pops, my mother's grandfather, was a locally-famed horse whisperer in his time. He could make a horse stand on its head, says my father. But what my mother also remembers was being forced to attend interminable gymkhanas, and perhaps that is what has been passed on. We stop to murmur to one of the four breeding stallions in the colony, Fenland Eric, a boisterous giant apparently given to biting if offered half a chance. Most males are gelded, as stallions are hard work. There are now only twelve in all of the UK. Some fields are full of nervous racehorses, which the Trust takes for rehabilitation. Uptight and antisocial, they are not used to living in a paddock. But here they do learn something about calmness when mixed with the Suffolks.

In other paddocks are geldings, breeding mares and skittish youngsters that rush to the gate and peer over with their jewelled gaze, as Ronald Blythe has described it. The paddocks on these sandy soils are cropped tight: modern horse-rearing seems to turn the land into a monoculture. The chesnut horses with auburn manes and tails look distinguished, the land less so. It would not have been like this in the days of farm working when horses were stabled and fed hay, straw and oats brought from the fields. They got all the exercise they needed from working. The old horse boys, as they were called, would not have recognised this land either. They were expert botanists, using forty species of wild plants to care for their animals. Today, having forgotten this knowledge, we call these plants weeds. George Ewart Evans wrote in *The Horse and the Furrow* of fevers treated with agrimony or with apples stored until infested with antibiotic-carrying fungi; and of colds and coughs cured with feverfew, belladonna, meadow-rue and horehound. They used box to keep down sweat, and burdock, saffron, rosemary, fennel, juniper, tansy and mandrake for coat conditioning. I leave John to his search for petty cash, pass a group of prisoners lounging in the sun, and head downhill to the marshes and ridged banks of Shingle Street. I am told that the prison service is less interested in rehabilitation these days. Yet across the country, the cells are full. One day, John hopes there will be a sparkling visitor centre here based on the celebrated Suffolk trinity: Punches, Red Poll cattle, and black-faced Suffolk sheep.

Down by the sea, I rediscover the shell line. The figure and sun have departed. The spring sea kale is deep purple, and I sit on the ochre beach and eat an orange, tropical bright in this land of brown shingle and blue sky. Sea kale shoots were once a delicacy, the traditional practice being to kick up shingle onto the plants in spring to blanch the shoots. A yacht drifts by in this horizontal world, and a bell out at sea clangs in the fresh breeze. I walk the few miles south to the Bawdsey gun site, where the multi-storey tower with red-tile roof gazes across the lagoons and to the glittering sea beyond. Walking back along the wall, I come up behind two women walking their dog. As I pass they jump aside, and there is a shrew scrambling its way across the grass. I mention to them the hawks, and how lucky we are too: a shrew running over your foot is considered bad luck. Walking on, I then see the kestrel, 100 metres ahead, stationary, quivering. That shrew was fortunate, but this bird does not let me up close. Maybe it's the chocolate-brown dog which races ahead and back, and up and down either side of the wall, tail bobbing in the long marsh grasses. It finds itself behind a close-linked fence by a farm track, swerves away, takes a run up and lifts, legs neatly bent like a show-jumper. It sails over, and runs towards me, grinning. Inland, lapwings are displaying, wheezing and creaking in local dialect, the tone rising at the end of each sentence. "Peewit, peewit". Swooping, white-and-black wings, dropping fast, twisting laterally, rolling as they fall. Skylarks are in full flow, and out to sea the bell on the buoy still tolls.

I read the map incorrectly, and think I have made too little progress. I am fed up, and consider calling Brian to apologise for being late. I tramp on, check again, and then have to smile as I realise I am actually a mile further along. My spirits lift, and I walk past the great sweep of low-lying Boyton Marshes to the left, and swing inland and upriver, leaving Orford Ness off to the right. The spring tide at over 4 metres is lapping near the top of the sea wall here at Flybury Point – it feels like an awful lot of water piled up to my right, and a lot of vulnerable low land to the left. Egrets glint against the dark clouds. The sun has long since gone, and the wind is blowing up. It was out in these reaches of the Butley River that Griff Rhys Jones found a temporary escape on one of his family's many holidays on cramped boats: "the solitude, the security, the sense of the journey made and the simplicity of the place".

After all that, I arrive three minutes early at the ferry. How emotions can distract from rational thinking: there never was a problem. I see Brian's shock of white hair from a distance, and by the time I've arrived, he's been across and back to check the far-side landing. There has been a ferry here since the 1600s, linking the western shore's mighty Augustinian priory to its lands on the east. In the course of this year, the eleven volunteer ferrymen will carry nearly 700 people across. Today, we embark and he rows, and somehow our conversation meanders until I find out that he was an architect who 40 years ago worked on the Wivenhoe campus of my university. He designed the very building in which I work. We laugh in amazement. I agree to write a piece on my walk for their newsletter. He dips the oars and we glide across the deep river, and then over the flooded saltings to land directly on the river wall. The jetty is underwater, far out in the main river.

I turn upstream towards the Druids' forest of ancient oaks at Staverton Thicks, before cutting inland to scuff sand on farm lanes lined with pines. A long straight road leads away from Gedgrave Marshes towards Orford. It is narrow with high hedges, forcing me to press into the vegetation to let vehicles pass. A battered blue car appears, the woman looking fiercely ahead, a child beside her crying silently. A library van squeezes by. I feel a bite on my cheek, slap away an insect, and there is a spray of blood. My hand is bright red, my cheek too. I'm astonished by the blood. Perhaps it's not mine and the blood has been sucked from another person or animal? I smear it away. In quiet Orford, I wait for the King's Head to open, and the day ends with a long bath in a room overlooking the flint church, and then careful study of the map for the next day's route to Aldeburgh and beyond.

I would be drawn back to quiet self-contained Aldeburgh with its shingle shore several times, and approach it by car along the straight spoke of a road, inland to periphery. This is the distinctiveness of many coastal settlements: they are termini. Unless, of course, you walk along the coast. On my birthday morning, I stand on the shingle shore and look out to sea before sunrise. I plan to go inland, and travel upriver to meet fresh marshes and farmland. On my way here, I saw the dark sky invaded first by bruised purple and then by lava flows of orange above a cloud bank on the horizon. It is half light as I arrive, and then the clouds suddenly disperse and the sun itself draws up from the sea, and bathes both the water and land with a salmon glow. In front of the old Slaughden windmill, I pick up a smooth circular stone glistening with salt water. Might this be the one Richard Long carried back from Aberystwyth? The sea continues its gentle grasp and rattle on the stones. "There is no sea like the Aldeburgh sea," said Edward Fitzgerald. "It talks to me."

Slaughden is Aldeburgh's lost twin. It prospered for hundreds of years as a fishing village, but is gone now except for boatyard and old windmill. Paddle steamers used to stop here. Well, not here, but somewhere out there in the sparkling sea. It's tranquil today, but in the late 1770s, George Crabbe watched a storm crash upon the village, the waves breaching the sea defences, and washing away eleven houses. In 1910, William Dutt saw here a "dilapidated sea-threatened cluster of cottages bordering a primitive quay and grouped around an ancient inn with a large bone of a whale suspended over its front door". In their last days the fishermen had to open the front and back doors of their cottages to let the tide run right through. The last person born in Slaughden was Ron Ashford in 1922, and his family's cottage,

Aldeburgh at dawn

the aptly named The Hazard, was swamped by shingle in 1926. The Three Mariners Inn was last to go at the end of the 1920s.

Slaughden had been famous for its cod boats, which also travelled out from Colchester, Wivenhoe, Brightlingsea, Lowestoft and Southwold to Iceland and the Faroes. They used long lines to catch the cod, which were baited with whelks bought from Leigh and Harwich. Until the eighteenth century the cod were salted, then well smacks were invented. These held a well of sea water to keep both whelks and cod alive, and when full could take 150-200 score of cod. On arrival back in port after months away, the fishermen used a short heavy club to kill the fish, hence becoming known as cod bangers. They are vividly brought to life in the gritty poetry of Crabbe and the later work of Benjamin Britten.

Crabbe was born in Aldeburgh in 1754, and his most famous works are *The Village* (1783), *The Parish Register* (1807) and *The Borough* (1810). In the preface to *The Borough*, Crabbe denied he had any particular place in mind, but it clearly owed much to Aldeburgh. His cruel fisherman Peter Grimes cast a bleak coldness on Aldeburgh itself. It is hard to see the town as anything but genteel today, but at the time life would have been harsh for many residents, especially those without land or boat.

Grimes is isolated from the people of the town, but not from the land or sea. He hears the clanging goldeneye, but never the concerns of neighbours and fellow fishermen about his treatment of apprentices. The rhythms of the place speak loud:

That winding streamlet, limpid, lingering slow,
Where the reeds whisper when the zephyrs blow;
With ceaseless motion comes and goes the tide,
Flowing, it fills the channel vast and wide;
Then back to sea, with strong majestic sweep
It rolls, in ebb yet terrible and deep;

Grimes kills one apprentice by kicking him into the fish well:

One night it chanced he fell
From the boat's mast and perished in her well,
Where fish were living kept, and where the boy
(So reasoned men) could not himself destroy.

Eventually Grimes is punished, and prevented from taking on more apprentices, and finally the ague of the marshes catches up with him:

Cold nervous tremblings shook his sturdy frame,
And strange disease — he couldn't say the name.

The river wall takes me inland into the Town Marshes. The still water of the Alde is glassy blue as the sun lifts further, not terrible and deep this morning. A grey mist clings to boats, and a short-eared owl appears from the long grass, lazily flapping, gentle beat on beat, and slips back into an area of wet reedbed. Behind me the town rises above the misty marshes, giving Aldeburgh a special poise. At his boatyard, I meet Bryan Upson to talk about the traditions of wildfowling and fishing, and later go on to see the lifeboatmen at the new station on the shingle. I stop at the Bru', the Brudenell Hotel, and sit in the sunlight shining from blue sky and glittering off the sea, the water molten and flickering with Brownian motion. At other tables, there's chatter, people meeting for the first time for, simply ages, isn't it? Gulls mew above the seaside quiet. The lanyard plays against a flagpole, snick, snick, snick.

I head back inland again to walk the upper Alde at the Maltings of Snape, where windows open to autumnal warmth emit the sounds of singing voices, pianos, cellos. Britten and Pears started the music festival in 1948, and Aldeburgh acquired national renown on the later opening of the new concert hall at Snape. The Maltings had been sited by the river both to land barley and ship the malted mash out to the breweries of London. I am here for the reedbeds and river path flowing towards Iken and the coast. A green line on the map looks like a good short cut, but I end up facing an expanse of mud across a broken section of river wall. Go back, it commands. I retrace, and take the longer and more sensible route past a line of dead oaks, 3 or 400 years old, twisted and white, having long since lost their living bark. Salt has done for them. Near by is a line of poplars giving off a papery rustle and their delicious sweet odour of fallen leaves. I gather one yellow leaf pocked with dark spots of decay, and press it in my notebook.

More oaks line the river wall towards Iken Cliff, dead perhaps since the 1953 floods. White trees, blanched as bones. In the hedgerows are lines of dead young elms. I prise off a piece of loose bark, and there are the radiating channels of the bark beetles. The bark crumbles into red dust. Along the cliff shore there are signs of erosion as these sandy cliffs tumble into the muds. There was once a huge warren here at Iken, producing tens of thousands of rabbits each year. Today, dozens of curlew, Ted Hughes' wet-footed god of the horizons, burble out on the muds and I continue on towards St Botolph's Church and nearby Troublesome Reach. Every river has one, say sailors. I look across to the north bank of the river. In the distant reeds is moored a wildfowler's grey punt.

A sandy lane takes me past banks of golden gorse ablaze with flower, and to a house with a glorious view of the river and its curving reaches flanked by marshes and reedbeds. I've arrived at Bryan Upson's home just after midday on this winter Sunday, and the sun is shining and for a while it feels more like spring than the depths of January. We scramble into full waders, pull on hats and gloves, and drive through the woods to Hazelwood Marsh. By a boarded gatehouse in a glade of dripping snowdrops we find the grey punt, pulled up in the swaying reeds. The sun glitters on the low water, and the mud sucks at our boots. Bryan built the punt wider than normal for extra stability, but it still only draws a few inches and is thus perfectly suited for gliding over mudflats. We stretch the camouflage netting over the boat, and load the decoy ducks, gun, rucksacks and Thermos flasks. The boat is launched into Ham Creek, and I walk alongside then clamber aboard as we paddle out of the reeds and into the wider Alde. Behind us is dense oak woodland, and to the left sandy cliffs topped with ragged red-barked pines. Ahead are the saltings

Punt on the Alde

fringing the sea wall, ideal for hiding from the ducks. Across the open water to the west, we can see Iken church tower on its vantage point.

It has not been a great winter for wildfowling. The worse the weather, the better the punt gunning. Sleet, snow and wild winds, preferably, as all these help to hide the boat, and also push the birds away from the centre of the river to the sides where they are more likely to be in range. Today, we paddle the punt across to the saltings, and set out the decoys. Seventy years ago, Julian Tennyson described these waters and his town of Aldeburgh in *Suffolk Scene*. He thought that restful, old and friendly Iken was the "loveliest part of the whole river". At the time of writing, "there was hardly a day in the whole year when fish were not landed", but all fishermen were wildfowlers too. When the wigeon arrived, "every man on the Suffolk coast who owned any kind of gun was up and listening half the night". At night too, the boom of punt guns could be heard in the town. Punt gunning, he said, was "an art in itself, a distinct and separate form of sport, which required the practice and study of a lifetime". One old punt-gun technique was to fire the gun downwards at shingle, so that stones would kill more than the shot. Tennyson would write no more books after *Suffolk Scene*, for he died in Burma during the war. "I have been out four nights in seven without so much as getting a shot," he wrote. "That didn't matter; it was the beauty of the river at night, the scents and sounds of the marshes, the treacherous slime of the mud, the tremendous sense of intimacy and loneliness."

Today the light is perfect for sitting and watching the gradual descent to dusk. Clouds of lapwing and plover fill the air over the reaches towards Iken. But there is too little wind, and the wigeon and teal do not lift and resettle. There is little reason for them to come over towards us, hidden as we are amongst the spiky yellow cordgrasses. Then two teal do drop into the point, and disappear into the marsh. The tide has lifted the punt off the mud, so we stalk them now, quietly and carefully, with splash and gurgle, water slapping on the boat, and there they are, swimming out, and then clutching at the air when they see us. Bryan brings up the gun, and the two shots of bismuth seem at first to have missed the birds. Now they are hurtling towards the wall. One goes over, the other splashes into the water. By the time we have paddled across, the second one has flapped up the wall and gone too, perhaps scuttling for a rabbit hole. It'll be a job for the two dogs later this evening.

And that is it. The only shots of the day. We listen for the whistling of the wigeon, see a barn owl swoop over the wall, watch cormorants beating across the river, look up sharply as two swans thud by. Across the way are babbling shell ducks, which were never, ever shot. There's a traditional recipe: prepare shell duck, put in oven with a brick, when ready, throw away duck and eat the brick. It's true for greenshank too, according to Arthur Patterson up at Breydon. Unpalatable, he said. Perhaps this helped Desmond Nethersole-Thompson in his remarkable sixteen-year study of greenshank at Rothiemurchus. Curlew, lapwing and redshank used to be shot for food, but not today. As the light continues to drift from bright to dark, there's a subtle shift in this soundscape of pipings and whistles and lapping water. Cold advances across the water. Then the sun drops further towards the horizon, and the sky begins to pink, and then flame to bright orange. A barge with a red sail drifts by Iken's now-silhouetted church, and reaches midstream, silent. Then we hear the pin-sharp voices of children, laughs and calls clear on this still air, carrying from a headland. To the east is Aldeburgh town, sunlit strip against the sea, the Martello tower, boatyard, yacht club and old windmill.

When we turn for home, the sun seems to pause behind Iken's low cliffs, slips away, and the cobalt of the sky rapidly descends. High clouds flare orange momentarily, and we paddle the grey punt back to the reeds. My feet are half-numb,

Aldeburgh from Freddie Cooper lifeboat

the rest of me relaxed and fulfilled. We return to drink tea by the flickering wood fire, distracting Bryan's wife and mother from the now muted football on the television. There's a warm glow in here filled with the lilting burr of Suffolk accents, and out there the river of sky is still visible through a window as darkness envelops the rest of the land.

The big bird clatters across the churning blue-grey sea, resolving from dark dot on the horizon to bright yellow. It heads for the body bobbing in the water. She's holding onto a flare, and the orange smoke pours across the water towards the shore twenty people deep with a rapt audience. Another squall whips across, and Aldeburgh disappears, then reappears. I can see the Martello tower and hotel to the south, and the blocks and white globe of Sizewell power station far to the north. Ahead, the coloured strip of the town lies behind the line of shingle. Now the RAF boys hover, rotors battering down the waves, and the winchman drops down and pulls Clare up to the *Sea King's* open door. Another circuit and it is above us, the water turning bottle green. Down comes the winchman again, trailing a line to discharge static. The lifeboat is heading steadily upwind, but dropping and rising, yawing and rolling in these vigorous waves. Relative to the helicopter we seem to be dead still, and he steps aboard with the ease of a cartoon character. Then down comes the casualty, delivered with precision, and the winchman is back on the line and up again. While they rescue a second body and deliver him to the tiny inshore boat, coxswain Lee takes the *Freddie Cooper* for a spin, roaring up to 16 knots, leaping across waves, through walls of water, powering along with guttural urgency, spray soaking our faces and hands, the only parts unprotected by our yellow suits and fur-lined boots.

Lifeboats say much about the coast and its people's refusal to let good voices be drowned. The RNLI has 233 stations in the UK, and their 4,800 mostly voluntary crew make more than 8,000 lifeboat launches a year. The running costs are £125 million, which is raised from the public who, amazingly, donate more than £300,000 daily. Each year, the number of people needing saving is increasing. Either we are getting more careless, or more people are taking to boats. Britain is a nation of pleasure craft: 450,000 of which are over 2½ metres in length. In all 3½ million people are engaged in some form of water sport. But that still leaves an awful lot of people who simply like the idea of lifeboats and their underlying role in community and national self-esteem. Since the RNLI was established in 1824, they have officially saved 137,000 lives.

It is easy to see how the lifeboat service emerged. Over the years many coastal communities set up shipwreck associations to collect and distribute the spoils of nearby wrecks. Some were involved in bringing ships onto dangerous sands and rocks in deliberate wrecking, more often they simply negotiated a rescue with the crew and shared the goods. There were substantial economic benefits as shipwrecks could be more profitable than fishing or labouring on a farm. Most towns had their salvage companies, some of which later evolved into lifeboat operations. Local boat builders designed fast cutters for use in all weathers. The first lighthouses of the coast were constructed at Orford Ness, Lowestoft and Winterton in the early 1600s, but proper lights were not widespread until the 1800s. For centuries, the night-time coast would have been dark from the sea, and individual storms could be devastating. One October storm in 1789 put forty vessels ashore between Southwold and Yarmouth. In February 1807, 144 bodies were washed up at Yarmouth alone. In December 1836, 100 boats were wrecked off the Colne. Off the Blyth at Southwold, not a single year passed without recorded loss of life in the late 1800s.

Now lifeboats are beacons of voluntary activity. The public pays to support a service that is very unlikely ever to help them personally. Coastal communities then supply volunteers who regularly put themselves at personal risk to save seafarers who often have not acted responsibly. They are rescued regardless of what actions got them into trouble and where they are from. When the RNLI rationalised their services in the 1960s, and closed a number of stations such as Sea Palling, Mundesley and Caister in Norfolk, the local communities decided that their lifeboats were too closely tied to their own sense of value and identity. They now raise the money to run their own services.

The sparkling and airy Aldeburgh lifeboat station sits right on the shingle on the outside of the sea wall. Nearby are fishermen's huts, boats drawn up on the beach, and the old lifeboat station that still houses the inshore boat. This station evolved from the Suffolk Shipwreck Association in 1851, and over the past 150 years the men and women of the town have saved 674 lives. They have been awarded fifteen medals for gallantry, and have suffered the loss of local crew, most notably in a terrible incident in 1899. The boat breached in great waves at the shore, was flung over, and six of the eighteen crew died amongst the shingle as the waves drove the boat deep into the stones. Up to the 1950s, all rescues were of trading and fishing vessels; now half are of yachts.

On the long walk, Gill and I strolled around the station, open to visitors, and admired the *Freddie Cooper*, built in 1993 and run with six crew. That day, we walked on, with miles still to cover. But a month later, I come back to meet the two retained crew, coxswain Lee Firmin and mechanic Steven Saint. The light glitters off the sea, and it seems so calm and benign. Aldeburgh is located towards the northern edge of the Thames coastguard's territory, which stretches from Kent to Southwold. The coastguard is the first to receive a distress call, and tries to contact local shipping to see if they can help. They then call the lifeboat station, and all of the local crew are alerted by pager. Once maroons were used to call them, booming from the shore, but well-wishers and the curious would clog the streets as they too made their way down to the beach. Until recently the maroons were still fired, but only several minutes after the pager messages went out. The community still wants to know when the boats go out. But there's now a problem. The manufacturing company in Essex has been taken over, and the maroon supply has stopped. A replacement mechanism will need to be found, as the signal to the town is too valuable a public relations tool to lose.

For many years it was a race. The first crew to arrive were guaranteed to be in the boat, but now the coxswain chooses. Those who go are pleased; those who stay are always disappointed. There are currently twenty voluntary boat crew on the roster, half of whom are from families who have lately moved to the town. There are another fourteen shore crew, and six on the management committee, and in all some sixty to seventy people involved in the whole operation. Twenty to thirty years ago, the crew tended to come from the same few families. Now, though, four of ten houses in the town are second homes, and lie empty for much of the year. The school has only fifty children, and many families can no longer afford to live in the town. Crewing the lifeboat may become increasingly difficult. Aldeburgh's identity is tied to this shingle shore, but those who buy second homes can end up ripping the heart out of the town. Lee, Steven and all the station are active in the community, keeping the lifeboat service at the front of people's minds. Later, when the 9 November high tide stalks the coast, they patrol the town as the water starts to overtop the sea and river walls, knocking people up and helping with evacuations. The inshore boat is ready for flooded street patrols, but in the end not needed. In December, they put up Christmas trees in the town, and a charity darts match raises money for the local hospital.

A tractor pulls the boat down the beach when the crew are aboard. The inshore boat is in the water inside five minutes, the larger all-weather one in seven to eight minutes. They go in bow first, straight through the waves. Nor'easterlies and sou'easterlies are feared as they drive waves into the side of the boat, threatening to overturn it. It is worse coming back, though only three times in the past four years has the all-weather boat not been able to return ashore. Out on the water, it's a life on adrenalin. You don't have a chance to think, says Lee. You are reacting to what is going on. Very occasionally, rescued skippers are unhappy. They don't want to be seen to have been rescued. Often a family member has called, concerned, and the lone yachtsman is embarrassed by the arrival of the lifeboat. Once they did arrive to find a yachtsman completely exhausted, but refusing all offers of help. The boat, though, stood off, and accompanied him back to safety.

Recent years have seen two dramatic rescues. In August 1996, a 10-metre yacht, the *Red House Lugger*, sent out a Mayday whilst in a force 11 north-easterly gale some 30 miles offshore. The *Freddie Cooper* was launched at 8.15 a.m., Ian Firmin the coxswain and Lee's father in command, with Steve as the mechanic. Both the Aldeburgh and Lowestoft lifeboats arrive at the yacht two hours later, but a huge wave almost carries the Lowestoft boat on top of the *Freddie*, which in turn is laid so far over that the cabin windows are submerged, and Lowestoft coxswain John Catchpole sees the starboard propeller clear of the water. In the 14-foot waves and 20-foot swell, towing is impossible, so Ian brings the *Freddie* alongside the yacht three times, and each time a crew member is grabbed and pulled to safety. The Lowestoft boat puts Bert Coleman on board the yacht and he lines up the final three crew, who are also grabbed. He also secures a tow. By now they are 35 miles east of Harwich and the two boats bring the yacht back to safe haven, arriving at seven in the evening. They refuel, and head home, the *Freddie* arriving at eight o'clock and the Lowestoft boat not until 11.30. Ian Firmin and John Catchpole were awarded the RNLI bronze medal, and Bert Coleman a thanks on vellum.

On a May Sunday in 2000 Lee is part of the crew that goes out to the Dutch yacht *Rose Bank*. It has had steering difficulties in a force 7, and the *Freddie* is launched at 10.38 a.m., again with Ian Firmin as coxswain. When they arrive at its reported location, there is no sign of the yacht. Eventually they find it 22 miles further away, and by now the gale is up to a force 11 and there are 20- to 30-feet waves. The four Dutch crew are completely exhausted and strapped in to keep them from falling overboard. Ian steams upwind, and a rocket is fired into the forestays, but the yacht broaches and the line has to be cut. From the wheelhouse, Lee can see that one moment they are towering above the yacht, and the next they are way down in a trough, gazing at the hull of the yacht up in the sky. A second attempt to secure a line catches in a guardrail, which is not strong enough. The skipper then decides to abandon ship, and the lifeboat comes alongside four times, lifeboatmen grabbing a crew member each time. Finally the skipper puts the helm hard over, runs across the deck and jumps onto the bucking *Freddie*. They arrive back in Aldeburgh by 3.15 p.m., the yacht's crew go by car to Harwich, and then take the ferry home across the North Sea. The yacht itself is later recovered by a fishing boat off Ramsgate. Ian is awarded another bronze medal for outstanding bravery and gallantry, and the Dutch crew come back to thank again the men who saved them, bringing presents and donations for the RNLI. Two years later, Lee took over as coxswain of the boat: his family have now been in charge for fourteen years.

When we hear of such rescues, we realise there is another reason why the RNLI and other local lifeboat services get our support. Like fire officers, ambulance staff and the police, they readily put themselves at risk for the public good. But in the case of lifeboats, there is no public money involved, and their rescues often occur somewhere beyond the horizon far away amongst giant waves, mostly beyond the reach of news cameras. And this must be why their own communities feel so proud of their crews' achievements, and why the streets are full when they know the boat is coming back in, pulled steadily up the beach, long stares on the crew's drawn faces as their families smile with relief.

I take the wheel, push forward the throttle, and ease the *Freddie* into the wind. I follow the bearing, and then swing to port when level with the southern edge of the town, around in a circuit and back up the shore to the north. Even at a modest pace you can feel the power and grace in this boat that can capsize and right itself, and has been involved in so many rescues. Earlier, when all the crew, shore team, town mayor and miscellaneous visitors were gathered in the mess room, coffees at hand, I got a sense of the difference between today's display for the summer carnival and the real thing. The Thames coastguard came on the radio, calling a yacht that had made a Mayday call. He called urgently but in reply there was only static. Everyone in the room stopped talking, eyes wide. Each leant forward, looking intently at the black box on the wall. Last night the inshore boat had gone out to this very yacht, but their offer of help was refused despite an engine overheating and rigging in poor order. The coastguard's calls continued, as did the static, but then it transpired that the yacht was now in Harwich's territory. If there was to be a shout, it was theirs. Within seconds, everyone was chatting again, joking, grittily commenting on the yacht's skipper.

On the *Freddie*, Paul is now detailed to watch me as I go forward on deck. The rule is that both hands should be used for holding at all times. I free one from time to time to snatch a picture as the deck bucks and falls. Paul talks of the crew's races to the station after the pagers go off, the occasional disappointment of not making the crew, and of the town pride in the lifeboats. Last night it was his daughter's birthday and they'd just pulled the corks on a couple of bottles when the pager pinged, and a few minutes later he was heading out to find the man who'd called in the Mayday and then decided he didn't want help. Today's volunteer bodies have been rescued. The *Freddie Cooper*, inshore boat and *Sea King* line up at the northern end of the shingle shore, and complete a perfectly synchronised fly and drive past. Everyone waves to the shore, and 1,000 or so people wave back. Lee turns the boat, instructs me to take a seat inside and hold tight, and accelerates directly at the shore. With a lurching jolt, the lifeboat crunches onto the shingle, lines are attached and she is cranked out of the water by the smoking tractor.

N
W
E
S
NORFOLK
SUFFOLK

Chapter 10

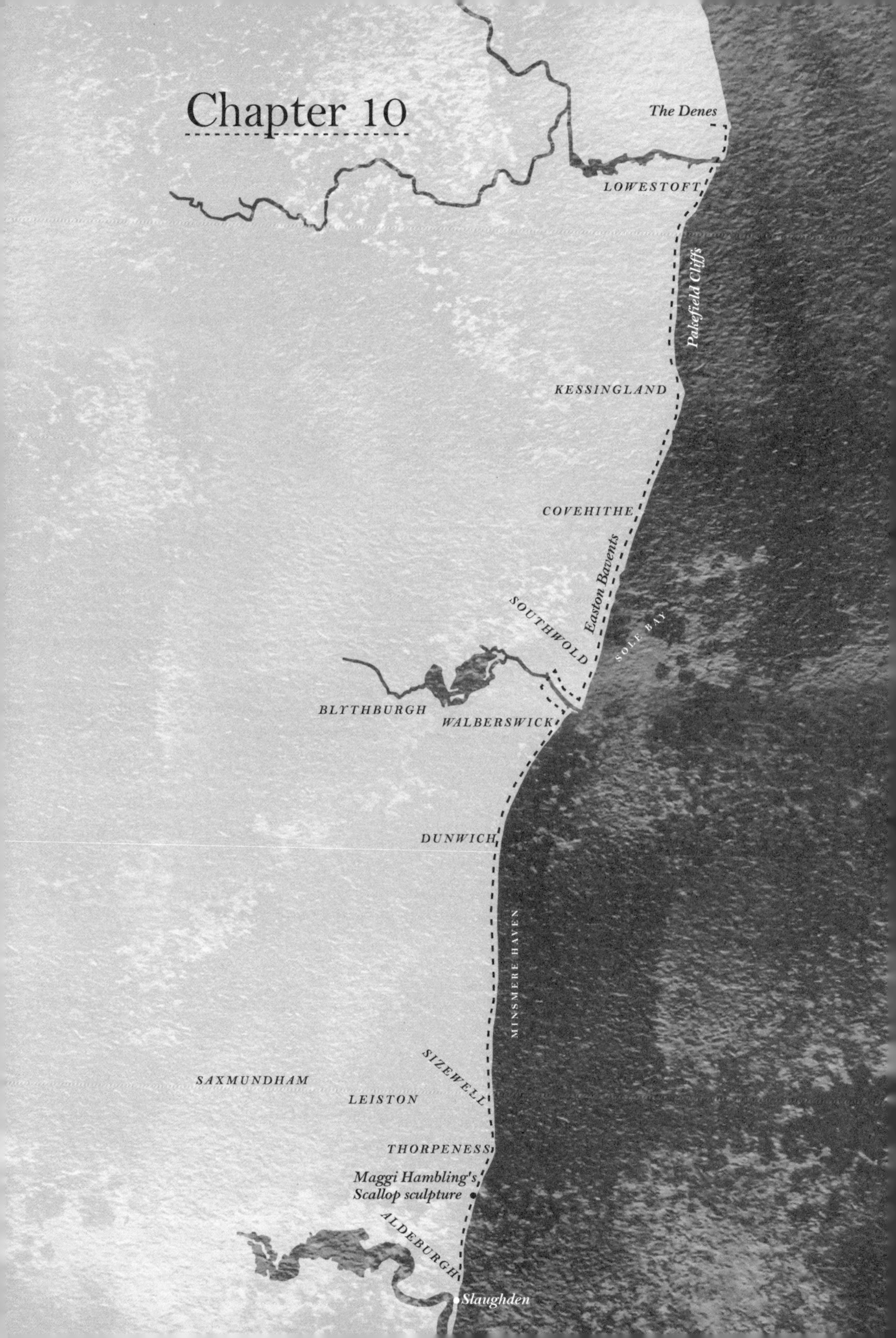
The Denes
LOWESTOFT
Pakefield Cliffs
KESSINGLAND
COVEHITHE
Easton Bavents
SOUTHWOLD
SOLE BAY
BLYTHBURGH
WALBERSWICK
DUNWICH
MINSMERE HAVEN
SIZEWELL
SAXMUNDHAM
LEISTON
THORPENESS
Maggi Hambling's
Scallop sculpture
ALDEBURGH
Slaughden

Maggi Hambling's Scallop, Aldeburgh

Chapter 10
Erosion & Memory

This coast was once wracked with poverty. Now it is the seaside. The few remaining working boats are drawn up on the beach, and tiny alleys lead from the front to the main street. At the top of Aldeburgh is the focus of a little local difficulty. Anchored in the shingle is Maggi Hambling's stainless steel *Scallop* sculpture, through which she punched the phrase, *I hear those voices that will not be drowned*, from Britten's *Peter Grimes*. Unveiled in 2003, it was immediately embroiled in sculpture wars. Some feel it spoils the beauty of a natural setting, even though from the sculpture itself you can see unnatural Thorpeness and Sizewell nuclear power station to the north. It's been daubed with graffiti: *THIS IS RUBBISH, MOVE IT*. Perhaps like Gormley's *Angel of the North* and beach men at Crosby, the disagreement will die down as the sculpture becomes associated with the place and a new identity grows. Today bicycles are propped on the back of the shell, and five children are climbing all over it, slipping, sliding, laughing. Two men stand with hands in pockets, watching and smiling.

Gill and I are heading for Dunwich, the capital of erosion, and walk towards Thorpeness, a strangely manufactured coastal village. As the communities of Slaughden and Dunwich were lost from this coast, others accreted. This entire area was bought in the 1910s by Scottish lawyer Glencairne Stuart Ogilvie, who built Tudor-style housing for his friends. A fantasy settlement emerged. The water tower was camouflaged as a residence, and became known as the House in the Clouds. The Meare which attracts many to feed swans and take to the boats was inspired by Ogilvie's friend J. M. Barrie, inventor of other fantasies in Peter Pan land. Thorpeness is still mostly owned by the Ogilvie family, and many of the roads are not public. On the northern edge, we find faded white letters painted on a concrete wall. *No admittance*, they say. *The cliff top is private.* The place is an odd mix of the welcoming and exclusive.

On the rise south of the village, we watch the swarming bees up here, and the scattered beach visitors down there on this coastal bulge called a ness. At Sizewell Hall, we find a dark and cool hollow-way roofed with evergreen holm oak, and circular apertures in the pebbly wall through which we glimpse the hall. We peer in, and spy people moving about, talking. They don't know we're here, but we don't know what they're up to either. At tiny Sizewell village, boats are pulled up on the broad beach too. But this is no isolated hamlet, for here are hundreds of hectares of land covered by power station buildings. Sizewell A is now closed, like Bradwell down on the Dengie, but reactor B is still in commission, covered by its improbably white dome. There are now plans for another reactor here.

The nuclear plant is surrounded by warrens and sand dunes, cropped tight and rich in plant diversity. Today almost no one eats rabbits: they can't compete with all that cheap chicken. It was by these Sizewell dunes that we once came as small children to play on the beach in front of the cranes and construction site. It was a day of sunshine recorded on flickering super-8 film, everyone looking impossibly young, skipping and jumping in the sand and waves, the colours bleached to light blues and pale reds. Looking at old photographs or film can be upsetting. We see ourselves as we once were, but then ask, as Griff Rhys Jones did, "When did this all happen, the time that went?" It is loss that besieges us. Beyond Sizewell tan Konik horses are grazing on the marshes, brought in by conservation organisations because of their liking for wetlands. Further on are Minsmere's marshes and woods, the jewel in the crown of bird reserves, full of rarities, including now booming bitterns. There were plans to import white-tailed sea eagles to the Suffolk coast from Poland in mid-2008, but then some objected: a sea eagle might take an endangered bittern or a lamb. The opportunity of grand spectacle appears to be lost. Would not those great predators bring more people to this coast?

Beyond sandy paths between concrete war defences, the vivid purples and yellows of Dunwich Heath rise up on the first great cliffs since Walton back in Essex. We climb to the terrace of white coastguards' cottages, and then swim out into the sea of heather and trefoil swarming with bees, to us nothing like the blasted heath and wild waste variously described from Dutt to Sebald. Beyond is Greyfriars Wood, where I creep to the very edge of the cliff and feel more than a twinge of dizzy fear looking down on the consuming sea. Out there lies Dunwich under the yeasty waves. We pop out of the dark wood into a meadow, and come upon the stone remains of the Franciscan Friary. Dunwich is one of 3,000 lost villages in England, and was once a substantial medieval port. Most villages were lost in the 1300s to 1500s,

Southwold from Easton Bavents

some because of the Plague, but more often the result of voracious landlords converting arable land into profitable sheepwalks. They are desolate and sad places now, but at least you can visit them, stand on house platforms and imagine a once vibrant community. Not so for those under the sea. At the time of the Norman Conquest Dunwich had six parish churches, two chapels and a Knights Templar church. It was granted a Royal Charter in 1199, and was home port to eighty large ships a century later. But a storm in January 1328 saw the first buildings disappear from the town. By 1650, only one church and a few houses remained, and in 1919 the tower of All Saints finally went over the cliff edge. Dunwich's dance with the sea was almost done. "Death and change, and darkness everlasting," said Swinburne. "This land of utter death." Henry James disagreed: "I defy anyone, at desolate, exquisite Dunwich, to be disappointed in anything." We're not going to be disappointed, for here is an inn, and the end of the walk.

To the north of the fine regency town of Southwold lies a band of crumbling golden sand cliffs called Easton Bavents. In the 1940s, my father and his cousins used to lie in the sweet hay meadows, listening to skylarks singing to the sea, and watch droves of bombers spiralling high into the sky as they prepared to fly across the water. To them, this felt like the front line, at the easternmost edge of Britain, the coast defended with tank traps, concrete blocks, pillboxes and Army posts.

Southwold lived by fishing, boat-building, wrecking and sea-trading. The salt works were as famous as those at Maldon for a couple of centuries, but are long gone. St Edmunds' flint church was not a wool church like so many others in the county, but a fish church built on herring and sprats. The 120-foot white lighthouse is the other skyline feature, its light visible for 17 miles out to sea. Like Slaughden, the town was famous for cod bangers who sailed north in March and returned home in August ready for the herring season. In the early 1900s there were 300 drifters landing herring into Southwold harbour. The borough was one of the first in the country to provide council housing. The beach huts, for which the town is so well known today, were scattered across the common between the town and its harbour during the war to prevent enemy planes from landing, and the beaches covered in barbed wire, steel spikes, concrete blocks and landmines. Local brewers Adnams say this is where the Dutch first brought hops to our shores, and so is home to the English pint of bitter. It may or may not be true, but the brewers have become one of the greenest businesses in the region, using super-thin bottles, recycled heat, herbicide-free hops, and a new distribution plant with sedum roof and walls of hemp. This, then, is our town, a still point in a turning world.

Meningitis invalided my grandfather, Father Tom as he was known, out of the RAF early in the war. With no source of income he came with his family from Lowestoft to Easton Bavents to live with his sister Violet, Vic, who ran the cliff-top dairy, and their three children, Harry, Peter and Mary. They were given 12-50, a polished mahogany railway carriage with sliding windows and a lean-to kitchen tacked on the back. It would eventually burn to the ground in the 1950s after a gas leak. Twelve-fifty had originally run on the Southwold railway that closed in the late 1920s after fifty years of sedate travel back and forth along the track to Halesworth, where it connected to trains for London. At its height, the railway carried 85,000 passengers a year, then competition with the bus led to its decline.

"All that is certain," W. G. Sebald observed in his *Rings of Saturn*, "is that night lasts far longer than day." His account of the Southwold railway centres on its links to China and the dark horrors of the Nanjing massacre of late 1937. Part of the story is that the Southwold train "had originally been built for the Emperor of

China", and that beneath the black paint of the carriages lay the outlines of imperial heraldic dragons. He did not succeed in finding out what happened to the locomotives and carriages. In fact the No. 1 loco, the Southwold, ran for only nine years before being resold to its makers and ending up on the Santa Marta railway in Colombia's banana plantations. The carriages were built in Bristol and intended for the Woosung Tramroad, but were never delivered to China. The real Chinese link was through W. G. Jackson, the main loco foreman, a colourful character who retired in 1916. He drove the first train to run in China in 1876 on the Woosung Tramroad and later built Shanghai Cottage in Southwold, where he and his wife held parties at which they dressed in traditional Chinese costumes and told great stories. It looks like it was Jackson who originated the China locomotive myth.

My father and his cousins were lucky. They were amongst the few children in Southwold not to be evacuated in the war. For much of this time, the school was closed and they spent their days ranging far across the cliffs, beaches and marshes. Food was always short, so wild produce was essential, as well as food pinched from the farm, especially the potatoes steamed for the pigs. "They were lovely," says my father. "I'd go there and grab one, snap the skin off and eat it with salt." A luxury lunch was turnips with jam; a luxury tea, a snared rabbit. The land between Easton Bavents and Easton Wood to the north was set aside for battlefield practice, and the soldiers stationed there loved having kids around. The boys spent their time collecting ammo, and making their own bombs from the cordite, from lime and battery acid, from dried peas in jam jars.

"We made our own enjoyment," says my father. "But they never caught us," declares cousin Peter Boggis of parents and police. The boys carried catapults everywhere, latter day hunter-gatherers, but also used them to smash the porcelain holders on telegraph poles. "We were hot shots," proudly states my father. Behind the farm, the marshes were cut off by a 10-foot-deep tank trap with steep sides. The boys made their way over the marsh to town by jumping from one concrete block to the next, knowing that mines filled the gaps. Their parents told them not to, but they did.

Planes etched themselves on the memories of the boys. In the first couple of days of the war, Peter remembers two steel grey Focke-Wulfs sweeping in through the mist, coming over the farm, and then racing back to sea. On another occasion, a bomber heading for the munitions factory at Beccles tried to bomb the farm, dropping bombs that overshot on the first run, and then undershot on the second. The house survived, and Peter saw the faces of the airmen as they flashed by. Later a doodlebug

Easton Bavents

flying across Southwold was blown up by cliff-top gunners, the explosion knocking out every window in the town. But it was the big bombers that stuck in the memory. In the mornings, the American Liberators and Fortresses would stack up over the cliffs, like so many spiralling rooks and jackdaws, gathering together from the 100 or so airfields in the region, and then would head as one for Germany. As they returned, shattered and bloodied, many limping in over the cliffs, so the Lancasters and Halifaxes of the RAF would be similarly stacking up for their night runs.

"We were the first to the crashes if we could be," says my father. They looked for the gyros and other equipment. Once they were first to a plane over by Walberswick and casually picked up an airman's helmet, finding a head inside. The worst tragedy they can remember was when one of two fully loaded Libs circling over Henham clipped the tail of the other. Both came down, only some of the crew managing to bail out. Firefighters from the town were quickly on the scene, and just as they seemed to have the fire under control, the 12-ton load of bombs went up, killing thirty-six firemen and aircrew. Today these meadows just north of Southwold look as if they are unchanged, but the truth is that the actual places where my father and his cousins lay, waiting for the bombers' return, are out at sea. When you stand on the cliffs at Easton, a wide landscape with few trees but dense hedgerows fringed by the churning North Sea and the town of Southwold in the distance, you feel the loss.

Peter is well into his 70s, and a few years ago decided that enough was enough. Some 200 metres of cliffs had been washed away since the war, and eleven houses had toppled, trees and hedges slipped, and lanes crumbled. The North Sea pounds these cliffs year on year. Like all sailors, fishermen and lifeboatmen of the region, everyone fears a north-easterly gale on a spring tide. "It's a little bit hairy living in winter on these tops," says Peter.

A seafarer and civil engineer, as well as long-time sea watcher, he understands these cliffs. Controversially, he paid to bring in 250,000 tonnes of waste material dug out from a new bypass in Lowestoft, and had it deposited in front of the cliffs. Some local people didn't like all the lorries; others said the spoil ruined the view. Statutory agencies slapped an SSSI designation on the cliffs to stop any further action, saying there were important fossils here, and that the cliffs should be permitted to erode. Nearby groynes were shortened from 70 to 20 metres, so now the sea is scouring away Peter's home-made sea defences. For the authorities what counts are models, calculations and regulations. Peter is softly-spoken and not at all angry. He's now spent ten years working on defending these cliffs. He gently states, "Thass a reg'lar mess, but we've not lost any cliff since we put it in."

He's called Canute by the press, ever ready for a cliché. Peter's actions pose questions. Should local people be able to take action to protect their own stretch of coastline? How should distant environmental and conservation agencies respond to their needs? It is clear that these cliffs are now temporarily stable. "I don't wish any inconvenience at all," Peter declares, he just wants to carry on living here. Whenever I hear the song of skylarks, I think of Easton Bavents, and the boys in the hay meadows. This is a place full of memories, and when it goes under the waves, so those memories will be threatened too.

Who shall be protected from the sea, and who not? In early 2008, controversy struck a government Minister, whose family seat is on the north Dengie at Stansgate, where the previous autumn the Environment Agency had spent a substantial sum repairing the concrete sea walls. At about the same time, the Agency decided that it would cost £40 million to maintain the sea defences south of Southwold along the Blythe Estuary. As these protect only six properties, they said, the plan was now to let the sea in. The whole of the estuary and harbour would be substantially affected unless local people spent their own money on sea defences. But then the political tide turned towards the people in 2008 when an independent inspector's report was accepted. Mr Kenneth Smith concluded that the special scientific interest at Easton Bavents was best served by protecting against erosion, not allowing it to continue. He also decided that Natural England's plan to force erosion on occupiers was an "unnecessary and disproportionate interference with human rights". The letter further stated that "Natural England's decision to refuse consent, therefore, is quashed and it is directed ... to recharge and maintain the sacrificial sea defence ... at Easton Bavents." Later in the year, a High Court judge confirmed the ruling, stating that the intent was the destruction of the soft sea defences and the rapid erosion of the cliffs. Peter says, "Inconvenient or not to bureaucracy, the defence of the coast should not be walked away from."

- - - - - - - - - - - - - - - -

The dog days of August bring grey-black clouds heavy with rain. I walk with Rob Macfarlane and Jeremy Millar south from Southwold to Dunwich to follow some of Sebald's route. Rob has stayed overnight on the cliffs and Jeremy and I meet him at the single-roomed Sailor's Reading Room. It's crowded with models of ships in glass cases, photos of famous fishermen, and figureheads high on the walls. We walk past the kindergarten I briefly attended, and then to Gun Hill with its line of five black cannons facing the sea. Each is smooth and shiny on top where generations of children have climbed up and inched forward to the end. I did it nearly a half-century ago; today a group of kids are doing the same. Beyond is Ferry Road with its eclectic and wooden houses, each one different, the marshes behind and towering dunes out front. At the harbour out, we stop at one of the creosoted wooden buildings along Blackshore for fish and chips, the national seaside meal. There are more than 8,000 fish and chip shops in the country, and 150 million meals are bought every year, so they say. We make do with three, but it's still far too much to eat. We walk to the Bailey Bridge across the Blyth to compare the scene today with the photograph in Sebald's book. This is the bridge I used to be terrified of crossing as a kid because of the wide gaps between the planks. Much later our RAF engineer Uncle Pete's ashes were scattered on the water here, and my grandmother's taken by boat out to sea.

The bark of withered elms crumbles in the green lanes of Walberswick. Jeremy tells how as a film-maker he visited the post-apocalyptic site for Tarkovsky's film

Stalker on the Jägala River near Tallinn. Here was the bleak Zone building used in the film, and when he went inside it looked exactly how it was pictured in the film. Later at home he was shocked when he watched the film again and found that the interior was filmed in a studio, and looked entirely different. We wonder at these mysteries of memory, and then return to this land. Crossing the Corporation Marshes we see a slightly raised island among the papery green reeds, and pushing through tough gorse we stumble upon a white line of bell wire leading us on. We come upon a lost world a few feet above this water world with mown paths, chairs, table, mist nets and speakers for a sound system. It's completely hidden from every angle, a secret site for bat or bird people. We stand on this gorse-enclosed space station, looking out. We want to stay, but Dunwich calls, and we tramp through the shingle to the north beach at the end of the cliffs, where Rob and I swim. We hobble across the shingle, and into the crunching stones of the mud-brown North Sea, which is as cold as ever. The ledge drops sharply away, and we splash out towards Bridge Street and the 1760s which is somewhere down there on the seabed. We had intended to swim to the 1500s but the flood tide rips us south and suddenly boats and people recede, and we have to swim aggressively at an angle to make the shore. We stand dripping in the wind, and then watch two youngsters fling themselves back and forth in the waves. We go in search of reward at the Ship Inn, dark and polished, seasoned but empty, then refreshed wander across the purple heather and onwards to Westleton.

My last day on the Suffolk coast has more appointments with memory. In the morning, after the guilty luxury of a cooked breakfast, my brother drives me back to the coast past beautiful Blythburgh church, where the legendary black dog once appeared, and drops me at Southwold pier. We talk of playing on the beach as kids, running down the slope from our house in the sunshine. I set off to the north, past Peter's cliffs. It is early Sunday morning, so I don't call. At rows of holes dug by sand martins in the cliff face, I realise that this will be the first time I have walked between Southwold and Lowestoft, though I've passed along the inland road many times. It is a curious coast, as arterial roads head to coastal settlements like spokes on a wheel, but there is no linking coast road. You can drive up and down the county and never see the sea.

The tides today are kind. They start low and allow me to get around cliff promontories, revealing the full sweep of shingle types and sands. Along here, Beaker pottery, barbed harpoons, needles, carvings and flint adzes have been found. On the beach is amber from the Scandinavian pine forests, and carnelian from volcanoes in Norway. But Easton village, including both St Nicholas' Church and St Margaret's Chapel, is in the sea. I arrive at Easton Broad, the first of a series of unique brackish lagoons along this part of the coast, called by Julian Tennyson "one of the quietest and most deserted places in Suffolk". Sea water enters these lagoons by underground percolation or overtopping during storms or on high spring tides. Incoming skeins of several hundred Canada and barnacle geese splash down in great chattering groups. Some whiffle to lose height and speed, dropping with chaotic splashes. As at Mundon and Snape there are petrified oaks along the broad's boundary, and silver birches perched on the cliff edge, soon to fall to the beach. The more saline Covehithe and Benacre Broads lie to the north, but Easton has dense stands of reeds fringing its edge.

The sandstone cliffs are bright orange and yellow with herringbone patterns of gravels. Covehithe was once a thriving medieval fishing village and now all that remains is the church of St Andrew sited within the ruined walls of an older church, a few hundred metres from the cliff edge. Since the 1800s half a kilometre of Covehithe lands have gone, another village and its memories all but wiped away. In this tiny place Bilious Bale was born. He wrote morality plays under Cromwell, became a bishop under Edward VI, and after a spell in hiding when Mary was on the throne became a canon at Canterbury. Benacre is the third and largest of this necklace of brackish broads, and harbours more geese and ducks, and a lone sentinel tree on the beach. It is all that remains of a once great forest in which James I hunted deer. Is this perhaps the money tree, made famous half a century ago for yielding 900 Roman coins in its roots? The OS map of 1837 shows a promontory here, about a half-mile out from this tree. A woman sits with her dog by the scoured and sculpted trunk, silhouetted against the sun over a liquid mercury sea. At home, I have a piece of bog oak from this forest, dark and dense, chopped as if for an ancient fence post.

At Rider Haggard's Kessingland the beach suddenly opens up and the cliffs retreat inland. Haggard bought Cliff Grange in 1900 and planted marram grass at the foot of the cliffs to reduce erosion. At the same time, Mayor George Staunton imported lupins, and between them they changed the ecology of the dunes, now known as marramels. The going is heavy through the fine sands and razor-sharp

Covehithe

grass. I climb to the cliff top, counting 56 steps only to find a locked gate. More private land on the seafront. I trudge back down. At Pakefield Cliffs by Crazy Mary's Hole, once frequented by a love-sick girl made mad by her lover's drowning, I watch three boys attacking the now dull ochre sandstone with sharp sticks, creating a great fan of loose material at the base. Above their heads, the top overhangs and the grass and thin soil will surely come down soon. No one says stop. A hundred houses have been lost at Pakefield in the last century, and their beachmen were once known as the roaring boys. They'd have said something. The sun beats down, the sea glitters and holidaymakers lie where there is sand, digging holes and holding hands. I decide to escape, and walk along the water's edge with my eyes closed, listening only to the sea sounds. Matthew Arnold wrote of these speaking shingles in his poem "Dover Beach":

"Where the sea meets the moon-blanched land
Listen! You hear the grating roar
Of pebbles which the waves draw back, and fling."

I hear the crump of a breaking larger wave, the rumble of stones pushed up the beach, the rattle of those drawn back, and the gentle crush of lapping water. This is a soundscape for meditation.

I open my eyes. A few fishing boats are pulled up at Kirkley and beyond them is a seaside church. I've seen pictures taken here of whiskered fishermen standing by their shods, houses made from upturned boats, with doors added and windows cut. None remain, though there are at least some boats. Just as the tide covers the best of the hard beach I reach the promenade. Up there on Cliff Road Benjamin Britten was born in 1913 and lived until he was 20. I come into Lowestoft and am reminded of the old Suffolk custom about always going out of the same door as you come in. I have never seen the town from this angle before. Up on the cliff top are the government fish labs, now closed, and my friend Paul's old flat with its glorious view, where we spent so many hours talking, writing and gazing at the sea.

Lucky Lowestoft. Ahead are the Claremont and South Piers and the wide sandy beach. The prom is lined with white concrete beach huts with bright wooden doors. Back in the 1970s we started coming to this beach every Christmas Day morning to swim. At first, just a few turned up, but then it became a town tradition. I remember the cold sunshine of the first occasion, our feet freezing in the sand. We lined up, and dashed into the water, and then rushed out again towards family and friends in their long coats, scarves and hats. Then we headed back for celebratory drinks, an excuse to warm up from the inside. Today, there is also a crowd, and they are applauding swimmers as they emerge from the water onto the shore. It's the finish line for a 1-kilometre swim between the piers, and many are wearing wetsuits. In the summer.

Lowestoft South looks vibrant in the sunshine. There are esplanade gardens, hanging baskets, putting greens, candyfloss and fish and chips, hotels and boarding houses. I'm secretly glad I've just missed the town's famous Air Festival, which can attract 300,000 visitors. In 2002, a Harrier crashed in the sea, the pilot's ejector seat firing him out to safety in front of the aghast crowd, one of whom was my niece knee-deep in the sea. The piers are less gaudy than some others I've seen, and despite officially being a deprived area, the town doesn't look downbeat. I cross the swing bridge, which breaks down as often today as it ever did, past the expanse of empty harbour that once jostled with fishing boats. There was no harbour here until 1830, when the first channel was cut to Lake Lothing. The original fishing village and community was up at the North Denes, but by the early 1900s there were over 500 drifters and trawlers registered to the harbour, and another 400 came down seasonally from Scotland. By the mid-twentieth century, almost all had gone. Lowestoft's full harbour was short-lived. It was an extraordinarily fast decline for a town rapidly built on the riches of herring, cod and mackerel. I walk up the high street, now pedestrianised. There are still many pubs, very few of which we could ever visit as they were the territory of feuding Teddy boys and fisher boys.

I head for the original fishing village, past Herring Fishery Score, Spurgeon Score and Rant Score, all tight lanes leading up to the high street on the hill. By the sea is the Birds Eye factory where I worked summers bringing in the peas to be graded, blanched and frozen. I sat in a hut by the sea wall, which we reckoned was the most easterly land-based workplace in the country. We phased the harvesting operations by radio, estimated the tonnage of lorries, calculated road distances and times to ensure that the factory was always at full capacity. I vividly remember the twelve-hour shifts seven days a week, and the utterly numbing effect of having no spare time, and the cycle rides home half-asleep. I pass more factory buildings, and then reach the Denes, a wide area of dunes and rough grassland covered with the remains of long wooden structures used for drying nets and making ropes. In 1904, Norfolk naturalist William Dutt said that "the fishermen's quarter that lies at the foot of the hanging gardens down from the high street is far more interesting than the South Beach and its throng of pleasure seekers." He found pebble-built cottages, wooden curing sheds and rope walks. Now the fishing families have gone and so have their cottages and huts.

Today, the sea is slapping the wall, and it's difficult to imagine this beach ever competing with the one to the south. I swing inland, past deserted and crumbling tennis courts hidden behind concrete and graffiti and up through the Sparrow's Nest gardens, a naval minesweeping centre in the war. At the High School I wait for my brother Chris to pick me up. I spent seven years travelling daily to and from this school, and it feels an appropriate place to finish. As I wait I decide to walk the Norfolk coast from west to east, and north to south, finishing at this very place. I've lived by the salt for ten days, spending every day with fluctuating tides and the light off the water. It has been a wet summer, but it didn't rain once on the entire journey from the Thames.

N
W
E
S
BRANCASTER BAY
HOLKHAM BAY
Overy
Marshes
BURNHAM
OVERY
BURNHAM
MARKET
HOLKHAM
HUNSTANTON
HEACHAM
THE
WASH
SNETTISHAM
SANDRINGHAM
Abbey
Farm
Vinegar
Middle
RIVER GREAT OUSE
KING'S LYNN

Chapter 11

Blakeney Point
STIFFKEY
BLAKENEY
CLEY NEXT THE SEA
SALTHOUSE
SHERINGHAM
CROMER
NORFOLK

Stephen and Peter Wortley

Chapter 11
Barrier Coast

North Norfolk and Fenland farmer Eric Wortley will be 98 years old on Armistice Day, but yesterday he saw a robin up close. It came in through the faded-green back door that gives directly into the kitchen, and perched on a chair. They looked at each other, unmoving, and then the bird flitted back out into the farmyard. When he was a boy, he collected birds' eggs, as did just about every rural boy at the time, but never those of a robin. It was bad luck. If a robin came into the house, his mother would say, "Tha'll be a death in the family." The best eggs were laid out on the sill by the white enamel sink, and the rest kept in a shoe box. The same sink in the same kitchen in which he sits today, for he was born in this Fenland house, and is feeling his years this cold afternoon. He sits in a straight-backed chair by the stove, the chimney leaking smoke into the room, and wonders about the passing of the years.

Eric was the oldest of twelve children, eight boys and four girls. But Mother died when he was 21, and Father just two years later, so he took on the farm. It was originally just 50 acres and now they have about 130, though it's still a modest operation by modern standards.

"I must ha' bin a rare good worker," he smiles. He was nimble then, he says, and he looked after all his brothers and sisters. Later, he raised a daughter and twin sons with his wife, now twenty years in the ground.

Today, Stephen and Peter are out in a field of black peat and sugar-beet, the rain coming in horizontally as they bring in the harvest ready for the short journey to the factory that belches steam on the horizon.

"We'd niver have survived," says Stephen, "if we'd had to pay for the beet to goo far. We're right lucky it's close."

The twins, one in red overalls, the other in grey, hoot with laughter at everything. One rides on an old red tractor, the other on the green harvester. No climate controlled environments for them, just a sack on the old metal seat.

Peter laughs, "We're all right boy," and rubs his earthy hands together. His mouth twists up on one side; Stephen's is an exact mirror image.

They are men of the land, perhaps a dying breed, and are in no way worldly. Their world is here, in this Fenland field, the bright-green leaves scattered over the ground, the roots of the beet crusted with inky soil. Grey clouds scud over vast skies, and yet more lashing rain strikes. Stephen lost a kidney recently, and is a bit poorly.

Peter says he may have to give up one of his.

"It'll be a perfect fit, sorta like", he says. They both laugh again.

This place is somehow parallel to the modern industrialised world. Hanging near the back door of the farmhouse is a long zinc bath which is brought in for use in front of the stove. Life seems bleak and unadorned, yet all of Eric's brothers and sisters lived to over 80 years of age – Brother Wes, Brother Dick, Sister Rene and the rest. Fenland people who survive the harsh early years of infancy live a long time. Their lives are of the farm, and of the wild too. Rabbits, hares, and partridges were staples.

"If we had nuthin' for dinner," says Eric of those days, "I'd goo an' git a pigeon."

Mole catching also fetched 3 shillings a skin, and an old boy came by bike to take the skins to the fur factory.

Eric worked the big hosses, feeding them at 6.30 and turning them out two hours later. They would plough an acre a day. Lunch was bread and cheese at midday, and dinner at about three o'clock. The horses were mostly shires, though Punches had a cleaner leg, he thought. Their first tractor was bought in 1946, he remembers, for £299, and it changed everything. Soon after, the horses started to go. There were so many flies in those days, he says: the horses and cows brought them in, and the kitchen would be full, clouds of 'em, in your ears and eyes. Out in the yard were chickens and pigs too. A proper farm.

Eric remembers the first herbicides to control charlock.

"Well, that was a miracle," he declares, "you'd be covered in white powder, white face, everywhere. But dew you know, it worked so well."

There was expertise too in the work. When thrashing corn, they set the shucks up in perfectly straight lines. Ploughing was the same.

"We were proper proud," he says.

He's never been to a pub, never smoked either. He used to play football for the village team. "You're captain, Eric," they said. But he broke his leg, falling from a shire horse on the farm and getting stuck beneath it. He was nine days in hospital

Eric Wortley

over in King's Lynn. The only other time he went to the sea in all his life was a trip to Hunstanton as a boy. The chapel hired a bus, and he remembers paddling his feet in the sea. But that was near ninety years ago. As a young man, he biked up to the village many a Saturday night to play snooker in the working men's club. Then he got married at 37. His wife had come up from Shooter's Hill in London for a week's holiday, they exchanged letters and eventually one of them put an x at the end.

He pauses at the memory.

"I always made a little profit," he declares. He's content with their yields, and does not aspire to match other farmers' achievements. "I dunno, they must be big acres to git so much."

But then he says, remembering the robin, "You come in a year, and I won't be 'ere".

This land, shaped, drained, hunted and farmed, sown and cropped, is better than when he started. It is also chock full of memories and a century of stories. It is firmly imprinted. And haunted. Eric and his Fenland family are indigenes of the marshes, a place that used to be one of the greatest wetlands of Western Europe. Until the 1600s, there were nearly 300,000 hectares of wetland south of the Wash.

Blakeney

Local people were hunters and gatherers who travelled by punt and on stilts, and lived by fishing, cutting willows, keeping geese, wildfowling and hunting. But officials labelled them as barbarous, lazy and beggarly, and national commissioners appointed Vermuyden to lead a great project to drain the marshes and turn them into productive croplands. The Fen men and women rose up and fought back over half a century, but by the early 1700s, the commons had been enclosed and those old ways all but gone. Today, small mixed farms like Eric's are equally endangered, and could easily go the same way.

North Norfolk has a long coast, but for stretches you can't see the sea. It's hidden behind a barrier of sand dunes, enormous beaches, pine woods, saltings and inned grazing marshes. From King's Lynn to Hunstanton's chalk cliffs, and then across to Sheringham's mud ones, tiny settlements are located at or on slight rises in the land, which puts most a mile or two away from where waves meet the shore. Their character is not defined by the immediacy of the sea, but by the many barriers that create an in-between land almost entirely devoid of permanent housing or settlement. You mostly have to travel to see the sea, though the worst of tides have brought it racing all the way to Brancaster, the Burnhams, Wells, Blakeney, Cley and Salthouse. This was not, though, always a coast where people and the sea were separated. Once Cley and Salthouse were two of the active northern Glaven ports, Blakeney only had a small spit curving across its river mouth, and Wells and the Burnhams also had ready access to the sea.

Just a small matter of 1,000 years ago, all the Glaven ports had active harbours and navigable ways from the sea inland to the settlements. For centuries they were the only safe havens between Great Yarmouth and Lynn around in the Wash. For these remote coastal communities, some of the best livelihoods came not from agriculture or marine harvesting, but from smuggling, wrecking and salvage. Wells acquired a particular reputation for wrecking, its residents nicknamed Wells Bitefingers after the practice, allegedly, of biting or cutting through the fingers of the drowned to remove rings. The inhabitants of each village had a nickname: the Cley Geese, Blakeney Bulldogs, Moreton Bodmen (snails) and Stiffkey Blues. In the Domesday Book, Blakeney was Esnuterle, and it took its current name from a Lynn fish merchant, John de Blakeney, to whom Henry III granted the manor of Snitterley, as it was called by then. Like Dunwich in Suffolk, Blakeney and Cley's peak days as ports were in the fourteenth and fifteenth centuries: Blakeney was the country's fourth largest port permitted to export horses, gold and silver. Great churches and friaries were built, and success seemed to be guaranteed forever.

However, by the 1600s and 1700s, the sea was regularly coming into the marshland and surrounding countryside, and Cley church was often flooded. Shingle banks and marshland disappeared as the coast retreated towards the towns. Although the inland marshes were drained with the help of Dutch engineers, the formal trade to the ports declined. As in other coastal settlements in Essex and Suffolk, this period was marked by the introduction of tariffs on trade and the subsequent rise of smuggling. By the 1720s, Defoe was commenting that trade was carried on with much less honesty than advantage, and that "the art of smuggling was so much in practice". Sometimes the revenue men captured whole boats of genever gin or intercepted loads of tobacco on the beaches, but mostly goods passed ashore with little interference from customs men employed largely from local communities. All the while, the shingle and sediments gradually migrated along the coast and the harbours and channels began to fill. Cley harbour was blocked by 1850, and a new quayside opened at Wells. Wells' success was further guaranteed by the new railway line from Fakenham that reached the port at the end of the 1850s, and the Earl of Leicester, owner of Holkham Hall, who financially backed the construction of a new embankment. Wells rose, and the other Glaven ports fell.

Tourism came late to this agricultural and marine coast. Paddle steamers and rail brought visitors to the resorts on the east-facing shores as far as Cromer. But these flinty towns, now so desirable that they are being hollowed out by second-home owners, were in no guidebooks. It was the wild nature of this coast that was to change matters. Naturalists F. W. Oliver and Sidney Long conducted scientific surveys over many years, and first Blakeney Point in 1912 then Scolt Head in 1923 were acquired by the National Trust. Cley Marshes were secured another decade later for the Norfolk Naturalists' Trust, and has the distinction of being the first county nature reserve in the country. The NNT spawned the Wildlife Trusts movement, which now collectively owns 84,000 hectares and manages more than 2,000 reserves. For this Norfolk coast wildlife-based tourism has become a major source of income.

One winter morning I drive to Holkham along the twisting coast road and past fields where Thomas Coke invented the Four Course Rotation, and held grand open days for thousands of farmers. The parkland laid out by Capability Brown occupies miles of countryside to the immediate west of Wells. Technically, it's Wells-next-the-Sea, but actually Wells is far from the sea and next-the-saltings. I walk through the dark pine forests opposite Holkham Park, planted by the third Earl of Leicester in the late nineteenth century, and come to a vast sandy beach. The tide is out, and a distant line of white waves sears the edge of the land. I inhale the resinous air, and am taken back to family holidays at Wells in the 1960s, staying in a caravan and playing on these beaches. I think we stayed in the forest, for I recall the pine needles, red squirrels and the oh so long walk to collect seawater by bucket.

I go out onto the sands, and there are razor shells everywhere, great drifts piled by wind and water. Nowhere else on these coasts of the East Angles and Saxons do they occur in such abundance. There are individual figures distant on the beach, some groups of two or three, a scampering dog. No one gets close enough to greet or smile; it's as if this land is too large for social interaction. I walk west towards Burnham Overy and its channel that once was navigable, but would barely take a rowing boat today, and then turn inland into the mix of embryonic, shifting and fixed sand dunes. It's a land of constant change. The embryonic are both prograding and eroding, and are stabilised mostly by couch grass. The white dunes are characterised by marram and lyme grass on the seaward side, but to landward they become grey with deep valleys and slopes carpeted with great banks of soft moss and *Cornicularia* lichen. It's distinctly warmer in these protected humid slacks compared with the open beach. From the summits there are fabulous views to inland Overy Marshes, where pink-footed geese whiffle into wide fleets of choppy water.

The rain arrives as forecast. It races across the exposed marshes, and now I know too that a whole day of walking is out of the question. This is proper rain, insistent and driving. I head back east towards the trees, and on the rabbit-cropped sward is a single blood-red scarlet pimpernel. Known as a weather-glass, it has five petals that close when it's wet. I check my *Rose* flora, and it says *fl. 6–8.* But this is December, not June to August. In the sweet forest, the habitat changes again. Where there is evergreen oak understorey, or deciduous leaves scattered on the ground, I can hear the pattering of raindrops from above. But where there are only stands of pine, the rain is absorbed by the needles, and the forest is disturbingly silent. Here too are thickets of green ferns, curious that in this north-facing landscape there are patches damp and warm enough to allow these plants to prosper. The forest floor rises and falls over old sand dunes and the path twists and turns. The sea has disappeared as has the blue sky. I am dripping wet. I find an inland-facing hide, watch the marsh for a while, and leave to pass percolation ponds linked by underground seawater seepage, in which the signs say there are endemic species of anemones and mysid shrimps. The inned marshes are now teeming with birds, and by the time I get back the car park is completely deserted.

Seasonal bird migrations also define this barrier coast. The mudflats, saltings, marshes and shingle banks are on a flyway for migrating geese, ducks and waders, which come in their hundreds of thousands from the far north as their winter advances. One mid-winter day, I rise early, the sky vast and cold, and head for Flitcham, a few miles inland from the north Norfolk marshes. The roads are empty save for ghostly barn owls and green-eyed deer. At Abbey Farm, we make our way to a pine copse on the hilltop, and settle by a hedge to wait for the first signs of dawn. Farmer Ed Cross leaves his sugar-beet tops in the fields after harvest, and we are here to watch the arrival of his pink-footed geese, widely known just as pinks. Twenty years ago they never came this far south, but now there can be a magnificent 2-300,000 birds each winter across Norfolk. These geese breed in

Iceland, eastern Greenland and western Svarlbard, and conservation organisations provide safe night roosts out on the marshes, mainly through predator control. By day, the geese want a good source of food, so the farmers, who these days receive too little for sugar beet, and get no subsidies for it either, forego some income by encouraging them to visit their fields. Without their positive management, there would be no geese here at all. They are pink-footed sugar-beet farmers.

Farming is so often a struggle to make ends meet, so why do these farmers do so much for wildlife? The first answer is that the pinks predominantly feed after the main crop is harvested. Unlike brents, they are not grass specialists so do not readily move into wheat fields. There may well be a net loss of nutrients, but then the geese are carrying some back from the marshes. There are some subsidies from government for farmers down the coast who manage grassland for brents, but not here.

Ed is justly proud of his pinks.

He says, "It tells me that this is not a barren land. We don't get paid for them. But we do get to see this stunning spectacle every day of winter."

Geese like these have touched others too. From his wooden hut at Baraboo, Aldo Leopold wrote that "geese proclaim the seasons" for Wisconsin. They arrived in his territory "gabbling to each sandbar as to a long-lost friend". Leopold also noticed that his flocks usually numbered six or multiples of six: "far more frequent than chance would dictate". In one essay, "Marshland Elegy", he describes first the "single silence [that] hangs from horizon to horizon", and then how "a glint of sun reveals the approach of a great echelon of birds". Geese and cranes made those marshes, conferring a "peculiar distinction" upon them, as they do here in Norfolk. Earlier in the same decade, Paul Gallico's *The Snow Goose* was published just after the retreat from Dunkirk, and was about geese on this east coast. Gallico, a New York journalist and war correspondent, lived at Salcombe in Devon, and was inspired by the story of a tamed pink in Sir Peter Scott's collection. He wrote about the wide marshes and saltings of Essex in a way that captivated the attention of millions. The outsider Philip lives in a lighthouse (as did Scott) and makes his living from the marshes and waters. A young girl, Rhayader, brings him the injured goose and together they nurse it back to health. The lone bird, separated from other snow geese, returns every year to the same place, and finally decides to stay permanently. Philip had long been rejected by the local community, yet joins the collective effort to bring the retreating soldiers back from the beaches of Dunkirk, where in the end he is lost at sea. Despite some discordance – the accents are wrong and there are no lighthouses on

Pink-footed geese

the marshes – the snow goose rightly entered popular imagination.

We face the bitter easterly wind as we wait for the geese from the west. They scare easily, and we are hiding. The first few need to come in, and then call to others that the field is good. The dark sky grades to mauve and pink and another barn owl appears, softly quartering the field boundary. Around us the liquid song of blackbirds and robins fills the air, and partridges chitter at roost. Then the advance parties of pinks appear, chevrons high up communicating mostly with high-pitched squeaks and looking like one of Scott's iconic paintings of geese against the dawn or sunset sky. Suddenly, their numbers fill the whole sky, as vast multi-family skeins arrive, sweeping past at first, then jinking back around the wood towards the field. Individuals split away and reform into strings and new groups. The pinks are chattering with a real purpose now. They do not seem worried, perhaps just sharing memories of the best fields. The whiffling behaviour is baffling, as some birds flick their wings to lose air, and drop metres in an instant, and others simply glide in and drop gently to the ground. The easterly sky is now a burning orange, and the birds form a bubbling mass of blue-grey among the beet.

They all lift. Just like that. No warning. Something has spooked them, and 12,000 birds rise up in one great outpouring of clattering sound. I hold my breath. The clamour of 20 or 30 tonnes of birds pours across the fields and woods in one great river of steely colour. The landscape rings with alarm and confusion, and all that quiet chatter is now forgotten. We stand in amazement, frozen fingers momentarily forgotten, as the giant flock flows up and around the pines on the hill. Soon order begins to emerge as the pinks form again into chevrons, and the cronking is gradually replaced with the higher-pitched neighbourly chatter. Eventually, they come back in again to the sugar beet. The green field reabsorbs them turning blue-grey again, and soon we can hear no more than contented murmuring. We creep around a couple of fields to get closer, and through the glasses make out one particular goose with a large collar ring. It's a male, CG (for collar grey) FJX, ringed in July 1999 in central Iceland, and since then recorded forty-three times in the UK in Loch Leven, Perth, Merseyside, Lancashire and across a dozen locations in Norfolk. What stories this old goose could tell.

A squeal of tyres, and two cars roar down the lane. We turn and see they're full of paparazzi with long lenses. In search of the young royals.

Then the pursuing security car screeches to a halt.

"Where are they?" demands one before his window is fully open. "And what are you up to?"

"I'm the tenant," soothes Ed.

Behind us, the geese have lifted again. Ed and I look at each other, and wonder how they could have missed the geese. We head back to his warm farmhouse kitchen. I pick up a pine cone and a piece of arrow-shaped flint on the way, and slip them into my pocket. At the farm, the lace tablecloth is laid. There's hot coffee and toast, and fresh Norfolk ham and mustard. Now it's our turn to eat. At the end of our winter, these birds will return to the far north to breed in May, and the cycle will begin again.

Four seasons in a day: hard frost, warm sunshine, scudding rain and piercing wind. Just as the geese are gathering up to leave the region, I come back to Wells with Gill to walk east and taste the saltings and marshes of the mid-coast region. A barn owl marks our arrival at Wells under a clear and cold sky. We start from the Buttlands, rimy open grass where archers once honed their skills, and head through the flint and lime-washed town to the harbour, past traditional butcher, baker, grocer and gun shop, and neighbouring charity and souvenir shops. Trestles fill the street, redolent of medieval times. Then the narrow streets burst open to the huge sky over an expanse of saltings. In the distance is the pink lifeboat station by the pines. The silent spray of the waves forms a white line between the blue sky and river mouth. In front of us, the fleeing ebb tide turns the boats on the water. Waders and geese pipe and gabble from out on the salt marsh.

We leave the rows of tiny harbour-front cottages and Smith's old mill and dockside gantry, now converted into flats. It is still close to freezing. The sticky clay sea wall towards Stiffkey is marked by a glossy lime-green-and-white line of flowering Alexanders, once a versatile vegetable and tonic. We will see only five species of plant in flower all day, indicating how spring has been delayed since its early promise. Above the green wall skylarks hover like angels. Out on the salt marsh that stretches to the northerly horizon we can now see flocks of brents and a few scattered greylags. They have paused on their journey north, and will soon be off for good. At low tide, the sea is 2½ miles away from the coast path. At very high tide, it washes against these walls, leaving a trash line of plant material, contaminated with plastic and jetsam from the modern world. To the right are ancient inland tracks leading to the marshes, Garden Drove and Cocklestrand Drove, down which livestock were formerly driven to graze the salty marshes. Do any farmers still do this?

By mid-morning summer has arrived and we're down to shirts. There are yet more saltings and then the great dunes at Stiffkey appear, locally called Meals. At the bottom of the sapphire sky there's still a white line of silent surf against the yellow-green saltings and towering cumulus clouds. Stiffkey was best known for its blues, or cockles, but later became notorious for the scandal of its rector, Harold Davidson, who helped prostitutes in London in the 1930s, and as a result was expelled from the Church. It's also one of those settlement names that seems to be designed to confuse visitors. Unlike Happisburgh, pronounced "Haisbro'", it is Stiffkey to locals while visitors often guess it should be "Stewkey". Beyond the meals (dunes) and freshes of Stiffkey is the distant tip of Blakeney Point. The western end of this great spit is home to a colony of 500 common and grey seals, a draw for tourists and an important revenue earner for the local economy. Despite the hotspot attractions, we mostly feel between places as we walk, and see only the occasional walker and dog out on the saltings. People in cars rarely stray far. But we do see some troubling signs. Paths are scarred by the tracks of quads and trail bikes, and where great puddles have been scoured so newer tracks have ploughed around them deep into the vulnerable saltings.

Sandy meals surround Morston quay and observation platform. Blakeney's St Nicholas Church tower peers out from a woodland promontory. Blakeney is flinty and sandy under a now watery grey sky. We swing sharply left beyond the harbour to Fresh Marshes, Blakeney Eye and the remains of a thirteenth century chapel out by the shingle. Across the spit is the Watch House, remote and lonely, once a coastguard station, then leased to the Girl Guides from the 1930s. When big weather comes, the waves crash clear over its roof. In front of us on the barrier mudflats of the Cley Channel are pairs of porcelain-and-black avocet, sweeping upturned black bills in the mud, and tottering on their high heels. They were absent for a century before returning to nest at Salthouse in the 1920s. Now they are a success story, with some 600 breeding pairs across east and southern England. A couple of enthusiastic Labradors scatter them into the saltings like papery butterflies.

The sea wall sweeps inland towards Cley's perfect brick windmill. It has four white sails, a bright cap and fantail and overlooks an area of swaying yellow reeds, entirely filling what was once the harbour. Cut reeds are stored in piles under tarpaulins, beside which are the cutters' machines. Cley was once a port, but is now just a village on the A149. It's another place that needs local instruction on pronunciation: it's "Cligh" not "Clay". We pass through narrow flint streets with quirky architecture, and as we walk back out to the sea somehow miss the Wildlife Trust's new visitor centre on a safe rise above the marshes. To the west is the route onto Blakeney's long spit, to the east miles of shingle stretching to the mud cliffs at Sheringham and Cromer. This is the only point all day that we see the sea up close. Curious for a coastal walk. Then autumn arrives: a dark cloud rushes across the sea and suddenly rain lashes down. We've been lucky to escape the scudding showers this afternoon, but now the many grades of toffee, pink, violet, charcoal, pearly and steely stones glisten, and I pick up a moonstone crusted with orange lichen.

The sea has gone again. We're behind a towering bank of shingle raised by bulldozers to protect of the marshes and towns. A couple of miles of stones follows, and we search out vegetated areas to make walking easier. All day, we've seen only the occasional group out walking, and only near the towns. Ahead is a lone walker. He's lingering, I think, waiting for us to catch him up. We turn instead down a footpath across the Salthouse Marshes into the village. The post office is closed, victim of the latest rationalisation, and sandbags are piled across the doorstep. There's a fish restaurant and a pub behind a stockaded brick-and-flint wall. A man

King's Lynn wildfowler's lodge

in a crumpled suit and pink shirt stumbles out into the suddenly bitter wind, and is joined by a woman in tight black outfit, both seemingly out of place as the wind rises a notch, and the sky darkens again. We wrap up tighter, and they stagger and shiver. We return to Wells, where the Crown Inn is warm and dry and its fine restaurant strangely quiet.

- - - - - - - - - - - - - - - -

There's a part of the East Anglian coast with quite different light and character. It's around the corner in the Wash, an embayment of 200 square miles of deeps and shifting sands, shallow inlets, bays and fringing saltings. The Norfolk coast here mainly looks west, where the sun meets water at the end of the day rather than the beginning. I start at Lynn early on a summer's morning with the land wreathed in mist. Once Bishop's Lynn belonged to the Bishop of Norwich until Henry VIII renamed it as King's Lynn in 1538. But local people don't defer and still, as I do, call it Lynn. It was once a mercantile heavyweight of northern Europe, as the centre of the Hanseatic League, the third largest port in Britain, seafishing centre and then a post-war overspill town for London. Now it's almost post-industrial. I park in the docks just beyond the last of a great fishing fleet, once including the only whalers of the region. Nearby is a hill of scrap metal by a rusting ship, and a giant grain silo that not only dominates this town's skyline but can also be seen from the north coast 20 miles away. Chemical works leak swirling clouds of steam, but there's no one to be seen.

The River Great Ouze is tamed and takes me north towards the Wash. It's a dead straight channel 4 miles long with steep sides and hellish mudbanks. A man appears, pushing a milk churn in a wheelbarrow, and he smiles knowingly but says nothing. According to the map I'm heading into lands with no distinguishing features. The roads north from Lynn run inland, and there are only occasional small tracks to farm buildings near the sea. No contours, so it's flat, and jagged lines mark former sea walls. There are no footpaths either. I had rung Jeff at the Environment Agency, and he said, "Just go." He walks the walls every year to check their integrity, and says it should be fine. I also speak to the head of the Estate office, and Marcus helpfully says, "Go; we'll tell the farmer." There are sensitivities on and near royal residences, but I plan to walk only on the sea wall. The footpath from Lynn does run for a couple of miles up to the Old West Sea Bank at Vinegar Middle, then peters out at a nature reserve of olive-green saltings and stands of cordgrass. Now the sun is out and the skylarks are singing as if the world is about to end.

Lone egrets, oystercatchers, and greylag geese are on the marsh. A cranky heron flaps away like a bundle of grey rags. These are the salty steppes that will fill with pinks on winter nights. At Vinegar Sluice, I climb over a gate to walk the outer sea wall rather than take an older inner one. I see cow pats on the wall. To the right several hundred woodies lift out of a yellow-brown rape field, their wings cracking together in agitation. They swirl, divide, sweep around and settle back into the crop. The job of the bird scarer was traditionally one of the loneliest in the countryside, but you can see why it was important. I walk towards Lynn Channel again, and then turn sharply north towards Bull Dog Sand and the open Wash. The walls here are straight with sharp angles following lines of reclamation projects rather than a natural lie of the land.

I see there are cattle on the wall ahead. Out on the outer saltings too, where spring tides can come up to the wall and there are treacherous muds, twisting dykes and open fleets. They're probably tough little blighters. The herd is a mix of black, roan and cream, and is an attractive sight on the green-and-lavender marsh under this wide blue sky. A Terrington farmer, Gavin Lane, confirms later that they are a suckler herd of Limousins, Angus and Charolais crosses, bred specifically for this life. He puts his out on the salt meadows in February, and they spend the summer picking up salt. They are then brought back in for winter, and sold off as salt-marsh beef. Gavin's grandfather reclaimed some of this coast, and the cattle's job was to graze off the salt for other animals and crops. Now there is added value in putting the animals outside the walls to give their meat something distinctive for the market. They're slow growers, but grazing also keeps down the grass, which in turn helps the nesting redshanks and oystercatchers. All very good.

A group is on the top of the wall. A large black individual stands with head turned towards me, ears flapped wide, shoulders hunched, as if awaiting a matador. I pause, then drop down on the outside of the wall, and walk along the saltings, and the animals keep pace ahead. I go up again, and they stop, turn and seem to say go on then. I try the inside of the wall and walk through banks of stinging nettles by a barbed-wire fence along a ditch. I'm ready to jump over and consider how long this would take and whether I could leap the water too. Then the herd stops, and I edge slowly and steadily past. It's a test. I look, and they look, but I keep moving. These are just domestic cattle, but a few weeks ago in Suffolk a woman was killed by cattle. It would be like being hit by a small car. I breathe deeply and climb carefully to the top of the sea wall. They drop down behind me and join the larger herd out on the saltings.

But then the black one starts high-stepping and jumping channels and suddenly the whole herd is running parallel to me. I pick up pace, and then start to scamper, looking left. They are still running, splashing and jostling. Ahead I see a gate on the wall, the escape. I make a run for it, and to my left they swerve towards the wall and start to charge. I can't win this race. Then unaccountably they simply stop. They are a few metres away and just watch as I climb over the gate and look back. The Vinegar Middle cattle: perhaps they are just lonely.

I breathe in, turn around and the land has changed. I am into wildfowlers' territory. The saltings here are half a mile wide, and all along the outside of the sea wall is a ribbon of gimcrack lodges and converted boats that are used by the Lynn wildfowlers in autumn and winter. It's a glorious sight. Remote, and not at all empty, this quarter. Each wooden lodge is different, developed organically depending presumably on the resources available. Some are on floats to cope with high tides. One is a boat on its side, the accommodation growing from it like a canker. They are numbered, and there are twenty-five or so in the 3 miles to Peter Black Sand. A dog rushes out of one and up the path, barking furiously at me, and a barrel-chested man in blue shirt calls her back across the ligger boards but doesn't look at me or acknowledge my wave of thanks. He disappears back into his huddle of wood and glass.

Above us the vault of the blue sky is crossed with condensation trails. To the left lie olive saltings and inside the walls golden wheat and barley extend to the horizon. The crops are nearly ready to harvest, but this will prove to be almost the last day of summer. The weather is about to turn so bad for six weeks that the harvest will end up being the latest for forty years. The night before the walk I had listened to a combine harvester roaring from over the hill, but today all is quiet. A barn owl rises out of the saltings, silently flies alongside me, then drops like a stone into the reeds on the freshwater side. The oystercatchers are more direct. They wheel around in the sunshine, vivid black-and-white with plasticine orange bills, shrilling and insistent, protecting their nests. A car creeps along a farm trail, and a man and a girl get out and investigate a patchwork mosaic of corn varieties in a field. They are some way off, but also don't look towards me. Inland, beyond the row of old sea walls, woodland rises up on a 60-metre ridge from Sandringham to Hunstanton, where the red-and-white chalk then falls into the sea.

The sun beats down. Where cattle have been on the wall, it is deep green with ryegrass. Other sections have fewer nutrients and are more diverse. One stretch is of shivering stands of wall barley, aflame in the sunlight, their bearded heads once used by children as clinging darts. I count 48 species of flowering plant today, and another 26 grasses. I reach 20 quickly, then the numbers leap again with Snettisham's unusual halophytes. I drink coffee lying in the long grass on the sea wall looking north at the distant strip of water and its silent ships moving towards Lynn. There are rabbit and hare tracks, and areas of flattened grass that would fit a couple of deer perfectly. I am wondering about sacred hares and superstitions when one pops up right in front of me, looks back, pauses briefly, and scampers ahead before zipping back into the deep grass. As I turn north towards Snettisham, I see another hare dead by the track inside the wall. I investigate: there are no obvious injuries, just a cluster of bluebottles crawling over its glazed eyes.

Signs at Snettisham reserve warn walkers off the beach. Inland is a line of ponds fringed with crumbling concrete, dug in the 1920s by a shingle company but looking like war defences. Along the beach are equidistant nesting terns and oystercatchers. The tide is right up and has come within a couple of metres of some of the nests. The

Heacham

birds don't look at all perturbed. The plants are now a mix of yellow horned-poppies with waxy petals and foot-long sickle-shaped capsules, vivid purple viper's-bugloss dotted with pink buds, frosted-green sea holly topped with their spiny neon-blue flowers. Then the wild grades into the settled, and these distinctive plants are left behind as the rest of the coast from here to beyond Hunstanton becomes increasingly populated. At Shepherd's Point a row of houses and caravans huddles right up by the hard sea wall. They have that distinctly individualised seaside style that you do not see inland, and are wedged between sea and the ponds, most with yawning west-facing windows. Surprisingly some don't have any seaward windows. Home-owners fuss in their gardens, others sit in chairs pulled up onto the wall. A man is using a strimmer to chop away bushes in front of his house, a buzzing wasp-like intrusion. A little further north, I can see Lodge Hill Farm perched inland on a hillside below a large woodland. In front runs the route of the now dismantled railway line. Somewhere along here, the 7.27 p.m. train to Lynn hit a wall of water in 1953, marking an early stage of the sea's victory over the land on that dreadful January night.

Today, the tide is retreating, leaving a strip of beach for holidaymakers. For 10

Hunstanton cliffs

miles, I have barely seen anyone. For the remaining 8, I will see thousands. At Heacham, opposite the wonderfully named Stubborn Sand, are great fleets of caravans, and the ribbon of beach is packed with families. Men in bright shorts and wrap-around sunglasses walk in animated conversation along the water's edge. Families gather together. Some squabble over nothing much, children hop over stony patches like they're hot coals, parents gaze to sea, hoping for more than a moment or two of peace. Here are scatterings of football shirts, but none from teams of this region. Heacham was home to get-away sailor John Rolfe, who across the water in Jamestown married a certain Pocahontas, who in turn came to a lonely end far from her home. But for most visitors this is just the seaside. Not far north of Heacham the numbers thin out again, and then at last I arrive at Hunstanton, which has been visible for hours today. In 1846, a new settlement was built south of the village of Old Hunstanton by Henry Le Strange, with hotels, a pier, esplanade gardens, a railway halt. For a while it was called Sunny Hunny.

Tourism might have been late coming to this coast, but you wouldn't know it today. I walk into full-scale resort chaos. The beach is packed, and there are pony rides, amusements, spinning carousels, ice-creams in hand and some dropped on the prom, grumpy people, harsh words, smiling people, people buried in books, fish and chips, and seagulls searching for a meal. Then here's the open green of the esplanade gardens, where there's a monument to those who died in 1953, and again packed with gasping families. Just before the coast swings from west- to north-facing, the famed striped cliffs appear. There are three distinct and dramatically different colours to these 18-metre cliffs. At the top is the white-and-grey Ferriby chalk, then a band of brick-red-and-pink Hunstanton chalk, also fossil-bearing, and beneath this the mud and rusty-brown Carstone. All are crumbling and eroding, though like the tree in the forest never when you are there. Scree fans spread across the beach, and great slabs of white and pink chalk lie in randomly scattered piles. There's a fringe of grass on the hilltop, indicating where 50 feet or so of open land remain before the first houses are located. There's no protection from the sea, no revetments. As the tide recedes, boulders covered daily by sea are exposed, bringing the bright green of seagrass into this landscape of saturated colours: blue sky, red rock, white rock, yellow sand, green seagrass.

Around the corner of the coast, the cliffs suddenly end and dunes reappear. There are no more cliffs until Sheringham. I walk through the crowds on the beach and then turn inland to call a taxi. There's no signal. At the Le Strange hotel, Fred and Anne on reception smile. The first words I've spoken all day, except to cattle. They go beyond the call of duty. I sit out front, cooking in the hot sun, and then I'm retracing six and a half hours of walking to the River Great Ouze in even greater inland heat. I walked lands that looked empty on the map, claimed them in some personal way. I just didn't expect to learn something about cattle too. Salt-beef sandwich, anyone?

CROMER
OVERSTRAND
TRIMINGHAM
MUNDESLEY
N
W
E
S
NORFOLK

Chapter 12

NORTH SEA
BACTON
HAPPISBURGH
HORSEY
HICKLING BROAD
Somerton
Staithe
POTTER
HEIGHAM
WEST
SOMERTON
WINTERTON-
ON-SEA
MARTHAM

West Somerton mill

Chapter 12
Mud Cliff & Marsh

In Poppyland, houses and churches fall into the sea. Following Jane Austen's recommendation in *Emma* about the healthy sea-bathing in Cromer, a middle-aged London drama critic began to take his annual holidays on the north-east Norfolk coast in the early 1880s. Clement Scott took a room in Sidestrand, and was entranced by the fishing villages, grand sea-facing hotels, smugglers' stories, the pier, and also fell for the miller's daughter. But it was the way the sea turned the solid land into the transient that shocked his urban sensitivities. You can stand on the mud cliffs, sea walls and dunes of this coast, and in almost every inland direction see church towers and windmills advancing steadily on the sea. Turn around, and out on the seaward side underwater are the remains of many once flourishing communities and their churches. Beyond Cromer is Shipden, a fishing village named in the Domesday Book that lost its church 300 years later, finally claimed by the sea in 1550. In 1858, the church had its revenge when it holed the steamer *Victoria* as she sailed over the ruins. A church on the seabed, a boat in the sky. Upside-down land.

Scott walked the cliff tops south-east of Cromer, and saw how Sidestrand's church was close to the cliff edge. The same was true of Eccles a few miles further south. He wrote a sentimental poem, "The Garden of Sleep", about these churches and their graves slipping into the sea. He named this land of eroding mud cliffs Poppyland after the poppies in both the corn fields and on the disturbed soils of the slumping cliffs. It was also the title of a later book he dedicated to the miller's daughter even though she had not, history tells, fallen for him. Poppyland became famous, and visitors flocked to see the churches before they toppled into the sea. And so the poppy entered national consciousness, setting the scene for its emergence as a poignant symbol in the First World War. They appeared in vast stands at Ypres, and medic John McCrae then wrote the poem "In Flanders Fields". In 1921, the British Legion established Poppy Day to mark the anniversary of Armistice Day, and all these decades later we still wear poppies in November.

In Flanders fields the poppies blow
Between the crosses, row on row…
We are the Dead. Short days ago
We lived, felt dawn, saw sunset glow,
Loved and were loved, and now we lie
In Flanders Fields.

"The Garden of Sleep" was set to music and sung for many years in music halls, and the tower of St Michael's Church of Sidestrand did then fall into the sea in 1916, twelve years after Scott's death. A century later, the cliffs are still eroding and homes are advancing fast towards their own gardens of sleep.

On the grass of the cliff, at the edge of the steep
God planted a garden – a garden of sleep!
`Neath the blue of the sky, in the green of the corn
It is there that the royal red poppies are born!
…
In my garden of sleep, where red poppies are spread,
I wait for the living, along with the dead!
For a tower in ruins stands o'er the deep,
At whose feet are green graves of dear women asleep!
…
In the dews of the deep!
Sleep, my Poppyland,
Sleep.

While the land and sea sleep under late October's full moon my brother Chris and I walk this same shore. North Norfolk is a land of the beyond tonight, barely peopled, it seems. Along Cromer's grand esplanade we see the first of many hotels and clubs that seem from the outside to be almost empty. A couple stand on their own in a huge chandeliered room. Another long bar is completely deserted, coloured lights strung outside. A deserted hotel has all indoor lights blazing. We search for a chip shop amongst the flint houses and heeled-over trees towards East Runton. At Bernie's, we order haddock and chips to go. On the wall is a poster for the legend of Chimera: there are Mongolian horsemen galloping about somewhere in north Norfolk. We sit outside and eat, looking across to the village stores, behind us Bernie's club, two more rooms harsh under ceiling fluorescents, one with nothing in it at all, set up for children to run around and bounce off coloured walls. The other's populated by men standing stiffly at the bar. This is a coastline of clubs as

Happisburgh church

well as pubs, it seems, likely because locals feel they need to escape, a way of keeping out the tourists when they are in-season.

Overstrand has piles of lobster pots and boats drawn up on the cliff top. We enter a nether world set between cliffs of Pleistocene sands and clays and the sea. Between the occasional lit houses and the dark deeps of the sea, and distant beacons of passing ships, no one will see us come and go. Indeed we will see no one until we get to Mundesley. Shutting-in time, as they call it up here, has come early tonight. We are by the crumbling cliffs beneath the young green of the winter corn. We had hoped for clear skies and the crystal silvery light from the full moon, but there's a layer of cloud. We only glimpse the moon once all night. Yet we can walk mostly without the torches, except when scrambling up and down the slippery cliffs.

An early defeat: the high tide forces us to turn up a cliff path and walk towards Sidestrand and along the cliff tops. At the edge of wheat fields chewed away at the cliff top, we peer over and see the slumped undercliffs and hear the crump of the sea some 60 to 70 metres below. Paths lead directly to the edge of the cliff. Tramlines in the fields also disappear into thin air, showing that this erosion is recent. We

Walcott groynes

wonder what the drivers of tractors and combine harvesters feel on these perilous tops. This is also home territory of Black Shuck, the great black dog with blazing red eyes that walks these cliffs, particularly under the full moon, they say. Anyone looking into its eyes either dies themselves shortly afterwards or experiences a death in the family. Black Shuck, though, may simply have been invented by smugglers to stop people going out. Legend says the smugglers would send out a small pony adorned with tattered cloth and a red lantern tied around its neck. Whether they made use of a convenient existing myth, or started it themselves, who can say. But then many lanes across the whole region have their own black dog and black cat stories, originating some say with the Vikings' Black Hound of Thor. There's Old Scarp in Yarmouth, and Owd Rugasan in inland villages, and of course, the Bungay Black Dog that appeared in August 1577 to the packed congregation at the church of St Mary. Even though we don't see Shuck tonight, this is still a mysterious coast. Somewhere west of here, there are singing sands where the wind produces the sound of lutes, and east of Sheringham are the shrieking pits of Aylmerson.

We step into space over the cliff edge. It's a long way down. We scramble over land thick with grass and moonscapes of sticky clay. There are patches of slippery freshwater seepages and sudden crevices. Eventually we jump with relief onto the wet sand where the sea was washing a couple of hours earlier. We turn south again past a mix of unprotected cliff faces and others behind linear revetments of angled railway sleepers. The waves rush at the crumbling concrete, sending explosions of froth high into the air. Now we feel vulnerable, walking between the looming cliffs and the sea. Ahead the gap between water and cliff narrows, and then after another mile we have to admit defeat again. The tide is still too high, and waves are slapping against the cliff face. We have to swim, or go back up. Now the clay is cloying, sucking at our shoes, and making upward progress a struggle. Above the high-tide line the climbing gradually becomes easier, although there are new landslips to negotiate. We scramble and pull, slip and jump on hands and knees, and then with one last haul are up on top again, breathing deeply, brushing off muddy hands, and setting off towards Trimingham, where there are more dark, empty streets and houses with bright deserted rooms. Televisions are flickering their ghostly light. Has everyone been whisked away?

A giant white golf ball appears. It's on a lawn behind a high fence lined with inward-facing sodium lighting. So much light, and again not a soul. Are there guards sitting inside watching us on CCTV, wondering why we're walking past this military installation at night? I wave at a place where a camera might be hidden. This Type 93 spinning radar unit has caused problems too, and in 2006, after much local concern, the RAF admitted it had been misaligned, causing car engines and lights to cut out, as drivers had claimed, and speedometers to swing wildly up to 150 mph. On the road to Mundesley we jump into hedges and verges as cars with distinctly working engines flash by. We walk past more nightclubs and hotels, a darkened housing development, and rows of houses in a mix of modern brick, traditional flint and seaside-resort styles. We talk about finding a pub, but arrive in the once popular Victorian resort of Mundesley at precisely eleven o'clock: closing time. But then a drunken woman in a dark coat struggles along the road, her hair and face wild, shouting at demons in another language. A young man with a blank face walks stiffly alongside and says nothing. Neither looks at us. Perhaps it wasn't so bad that we missed the pub. We stop instead at the war memorial on the deserted green. It's dedicated to the sailors and volunteers who cleared these seas of mines in the last war. We stand and read out their names, feeling they deserve to be spoken aloud.

Below is Mundesley's broad beach and we rest with our backs to the concrete prom.

"Do you think it right," I say, thinking of Edward Abbey in Glen Canyon before it was drowned, "that we're sitting out here when everyone else is locked down inside, answering late-night emails?"

"Yes," says my brother.

We walk on. There will be no more cliffs until Happisburgh, and on the shoreline 100 metres of sand and shingle are drawn by wind and water into striated patterns. In the distance, waves appear as first a tear in the dark sea, a luminous white line of foam spreading outwards in both directions, and then with a boom disappear. As we walk, the dark night begins to dissolve into another orange glow where a square kilometre of land is so drenched with light that it is bright enough to produce sharp shadows. Bacton gasworks, whose underwater pipes suck gas ashore from Siberia, and push it onwards to houses and factories. A gasworks wall for dreaming dreams, smelling the spring on the smoky wind. We're now in that dirty old town.

The shadows envelop Bacton itself, clouds still drifting across the moon, and Keswick and then Walcott gap too. The houses on the seafront are silhouetted against the light clouds. There is little to distinguish these settlements and no signs of life. We decide to stop a couple of miles short of Happisburgh, and wait on the sea wall opposite the village stores of Walcott, before heading back to Poppyland along empty twisting roads. Thirteen days later, the sea comes through at precisely this point where we had been waiting for the taxi. It smashes into houses, breaks windows, scatters street furniture, floods roads and leaves everyone relieved that the sea did not in the end rise quite enough to break into the great expanses of low-lying fields, marshes and broads beyond.

The view through the wooden bungalow's window is terrifying. And the weather is benign today. It's a few weeks later, and I am in Happisburgh to meet Malcolm Kerby. From the main part of the village, I turned down Beach Road, and entered threatened territory by the former lifeboat station. Here the houses have a temporary air, peeling paint, curtainless windows, crumbling garden walls, flowerbeds abandoned to weeds. One empty bungalow was recently valued at a single pound sterling. The compelling invitation of gravity hangs in the air. Small places with little political power need activists to bring their plight to a wider national audience. Malcolm is one. After he retired, he moved here and helped set up the local action group that is trying to persuade local and national government

Happisburgh

to come up with the money to replace the destroyed beach revetments. Seen in aerial photographs, the effect of the lost defences is compelling. Where the line is broken, the sea has entered, crashed against the mud cliffs, and eaten them away. Where the line has been maintained, the cliffs are whole. The earlier pictures from the mid-1990s show a line of houses running parallel to the cliff top. They no longer exist.

Where there are sand dunes to the south of the new bay, the Environment Agency takes responsibility for sea defence. If the sea came through anywhere between Eccles and Winterton, then salty water could race west towards Norwich and south to Yarmouth, and the Norfolk Broads would be no more. In the 1953 floods, 75,000 houses were flooded in Yarmouth. Where there are cliffs, however, sea defences are the responsibility of local authorities and national government. Malcolm shows me where that ghost row of houses once stood with their long gardens leading to the cliff tops. Now the sea is eating into Beach Road. The lifeboat was recently moved south to the dunes when the slipway, well, just slipped away. We can see the concrete chunks lying on the beach. Malcolm knows many politicians, and has had a few flaming rows. The local councillors are supportive, but national ones seem to follow the diktat of a new shoreline-management plan, which states that parts of the coast will be allowed to disappear while others are protected.

We walk across a wheat field and pace out the number of steps to a pillbox. We look at the way the cliff tops have slumped down into heaps of orange and yellow spoil, the way field tramlines now lead to danger. The pillbox is nearer than when he last measured. The pinch point is at the southern end of the bay, where the inland hills beyond the cliffs drop away to the low marshlands. If the sea comes through here, between the cliffs and dunes at the point where responsibilities shift between agencies, then it is in forever. Solitary elderly men walk stiff dogs along the tops, greeting Malcolm as they pass. Back at the end of Beach Road, we stand at the point where the road is sheared off, and peer into the windows of the last bungalow that looks down on the tangle of half-abandoned sea defences. For an island nation to let part of its land go seems short-sighted. And with the fourth largest economy in the world. Malcolm is off to chair the action group's AGM tonight, where they will hold more discussions about the endless struggle to stop this edge town being forgotten.

- - - - - - - - - - - - - - - -

When New Englander P. H. Emerson travelled to Norfolk and Suffolk in the 1880s, he brought his plate camera and left a unique record of people and place. He was a cousin of Ralph Waldo Emerson, and like him produced enduring icons of the land, pictures into which, as Ronald Blythe says, "a community stares in search of its soul". He moved to Southwold and spent a decade composing portraits of the overlooked men and women of farms and wetlands. His unique use of sharp, hard outlines for the subjects and blurred backgrounds for the landscapes contrasted with the practice of many contemporaries. Yet the cold, harsh images of his peasant subjects, like the reed-cutters with their scythes in *Coming Home from the Marshes*, endured far better than the efforts of many of his fellow photographers. In Emerson's short career, he captured images too of pike-fishermen, eel-babbers, wildfowlers, farm labourers, hay-makers, fishermen collecting water-lily flowers for bait, basket-makers, osier-peelers, poachers and ferrymen. They are the folk that inhabit Bensusan's stories of the Dengie, Patterson's Breydon Water, Tennyson's Aldeburgh, and Wentworth Day's Norfolk Broads and Essex marshes. Emerson did not seek neutral photographs, and annoyed rural elites who did not like his placing of the poor centre stage. As Blythe also points out, he published studies of these people so that the images "would help in the understanding of this particular region and add to the outcry against abuses".

Martham Broad

The broads and reeds of Emerson country are ghostly under the crisp silver light of another full moon. At the flinty terminus of Winterton in the quarter-light before a winter dawn, I walk through the sand dunes to watch the sun rise out of the sea. Somehow it seems right to stand. Behind me, between the tarred fishermen's huts, the moon sets as the sun rises, and now the dunes are rose pink, as is the froth of the back-lit waves as they deliver muffled thumps to the sand. The main village stops some way before these protective dunes, and here I am in a half-inhabited, half-wild zone. Two early dog-walkers are swallowed up on the vast beach, disappearing in the haze. Where once the Vikings gained a presumably bloody victory, Blood Hills are now covered with wind turbines, half of which stand broken in the breeze. By the base of the hills lies Burnley Hall and its ruined roofless chapel, from whose chancel an oak tree grows. An old woman with a wooden leg was buried here; the leg took root and the tree sprouted, they say. Beyond is Martham Broad, then Hickling, and to the north Horsey Mere and the Hundred Stream, once a route to the sea. But it was blocked long ago, and now all fresh water must drain away from the coast, behind its creaking fortress of dune and cliff.

Heigham Holmes mill

We've had a month of rain and are lucky. Today's sunshine brings an uplifting light to the yellow-and-brown landscape of reeds and water mills at Somerton Staithe. I meet Richard Starling, chairman of the recently formed Broads Reed and Sedge Cutters' Association, along with Keith and Mike, volunteers for the Norfolk Wildlife Trust. Richard is leathery and soft-spoken like all marshmen. He wears an olive-green wool hat and blue overalls that have seen some use. We climb aboard a flat reed cutter's skiff, outboard loaded, bags with thermos flasks and lunch stowed, and move gently away from the staithe and enter a dyke leading past the sentinel that is West Somerton drainage mill. Within minutes we are swallowed by a landscape defined by reeds and sedge, now golden in the morning light. A breeze ripples the glassy water, then all is still again. Water in the dykes, water among the reed beds, water in the fields. The mills now seem lonely: an old trick on the Norfolk marshes was to set the sails of windmills dead upright to indicate that the customs men of Yarmouth were on their way. This was James Wentworth Day's route as he travelled these broads in the 1930s. He wrote with great affection for the marshmen and their families, and seemed to know the place so well.

Roll-up clamped in mouth, Richard steers the skiff between the smooth mirrors of the north and south broads of Martham, and then up the Hundred Stream to a hidden reed-cutting base. To the north we can see a dark marsh harrier circling like a vulture over the savannah, rising, turning and falling again. From the reeds comes the chitter of bearded tits, and then the "seep-seeps" of rare Cetti's warblers. We moor the boat, watch the light change as the clouds roll back further, and then start to stack the bundles of reeds. There is such great demand for roofing reeds that some thatchers are now choosing to import them from Turkey and Poland. Richard wants the local traditions to continue, and his association is training four lads and a girl to enter the trade. It wasn't easy to get the money for the project: the government only seems to give support to deserving causes after they have completed a mountain of paperwork. Richard rolls his eyes at the memory. He'd rather be out here, in these great wetlands of fleets and ponds, water meadows, soughs and swamps, all oozing, dripping and leaking, the fresh behind the salt. But then again, reed-cutting almost died out at the turn of the millennium. And no management would mean scrub encroachment and a closing of this landscape.

A path takes us deep into the reedlands. We walk over ligger boards and sluices to another drainage mill, and past recently installed tubes that allow the otters to pass from one waterway to another. Water levels have to be constantly managed, and the crew set siphon pipes to run water from one reedbed up over the wall and down into nearby dykes. The boundary of this reserve links with the National Trust's Hickling Broad, and between the various conservation organisations large parts of these broads are being co-managed for both wildlife and reeds. In the sky are now two, three, no four marsh harriers, reedbed specialists. By the early twentieth century, they had been pushed almost to extinction by gamekeepers. They began their recovery here, the first breeding pair returning to Horsey in 1921, since when they have survived the threats of organochlorines in the 1950s and 1960s and now number 200 breeding females across the whole region. A Chinese water deer walks confidently in our direction, marks us, and then arches its back as it bounds across the grassy fields to another distant dyke. In the far distance is a herd of large red deer, and beneath our feet are the slot marks where their hooves have churned the Eelfleet riverwall as they passed. Cranes live here too, some thirty or so individuals, but they are in hiding today. We do, however, see a couple of thousand pink-footed geese grazing on the marshes having flown in earlier from their roosts. When something disturbs them, they lift, spilling across the sky, and right in the middle is a single

Reed cutters, Martham

white snow goose. It will winter here on its own, probably staying with the pinks for safety. All the geese pour across the reedbeds and sweep noisily into the sky, split into two groups, reform, and eventually sink into the marsh grasslands again.

Out here, there is a lot of sky reflected off the Broads. Though we are not far in miles from settlements, physical distance does not count for much. In the famously bitter winter of 1946-7, Horsey was completely cut off by snow, and local people had no bread, meat or tea. There is a constant threat of flood from the sea, and its "ceaseless growl of menace", as Wentworth Day described it. An old tradition when winters were cold was to put smooth leg bones of horses under a punt to act as skates, and then to move the punt on the ice towards the central open parts of the broads and meres. From these broads comes the gruesome story of a man who killed his sweetheart, dropped her weighted body in a creek, and then put out an eel-sett to catch the eels attracted to the body. There was a seasonal rhythm to life on the marshes, as Mark Cocker has recorded in *Crow Country*. Like those from the Fens, families from the Broads were hunter-gatherers, and some traditions survive. January was for cutting reeds and wildfowling, February for setting networks of trimmers to catch pike. In March, the focus was eels, and collecting early eggs, especially of lapwing. April saw the hunting of birds of the open marsh, especially ruff, and May godwits and tench from the waters. June was for marsh hay-making, and July for collecting flappers, young duck without flight feathers. By August, migrants started arriving with the first cold nights, and teal, curlew and snipe became targets. In the autumn and early winter months, wildfowl and geese were hunted, with local marshmen also acting as guides for sporting shoots.

Quiet returns to the satin sky after the clamour of the geese. We can hear again the trickling, gurgling, percolating water. A fish jumps to escape a pike and splashes down. We return to complete the stacking, each bundle 3 hands wide, some 25 inches. There is constant banter and laughter. Later, we climb aboard the skiff and head back towards the staithe, stopping first at a floating platform opposite the West Somerton mill. It has no sails, though I later find a picture in Wentworth Day's *Marshland Adventure* showing it with a full complement of four. Keith and Mike work to replace a damaged gatepost on the river wall. I saw the post and then help Richard dress the reed bundles, learning to comb out, knock up, and retie them. Before long a carefully stacked pile ready for thatchers emerges. Once a Minister came here to lend his support to reed-cutting, and on another occasion the Prince of Wales visited. The Minister was barricaded in by two groups campaigning against his views on fox-hunting and his department's stance on coastal erosion at nearby Happisburgh. Eventually, I leave Richard, Keith and Mike sitting in the boat on a coffee break in front of the brick mill framed against the expanse of reeds, and walk back through the slippery mud of the river wall to the damp and enclosed hamlet by the staithe. Richard will be here again tomorrow, and the day after.

N
W
E
S
NORFOLK

Chapter 13

APPISBURGH
ECCLES
SEA PALLING
WAXHAM
WINTERTON-ON-SEA
CALIFORNIA
CAISTER-ON-SEA
GREAT YARMOUTH
BREYDON WATER
THE BROADS
GORLESTON-ON-SEA
HOPTON ON SEA
BLUNDESTON
CORTON
LOWESTOFT

Norfolk coast

Chapter 13
Sandhills

We hold open a picture of the recent waves crashing over the sea wall, three times as high as the watching rescue workers, and then the wind snatches it and sends it spinning along the beach. It's two moons after that November storm surge, and I have returned to Walcott and the sandhills and seals of north-eastern Norfolk. Gill and I slip by the Yarmouth dog stadium and the grey estuary of salty Breydon Water, where some of the last hunter-gatherers of the region lived, skirt Hickling Broad and Horsey Mere and their swaying yellow reedbeds, and return to the bungalowland of Walcott. We struggle to find a place to park and settle for the grounds of the still boarded-up Fisherman's Restaurant. Along the low cliffs are signs of the damage to a row of eclectic bungalows. At one, an armchair sits in the ashes of a fire in the middle of a lawn; at others windows are boarded. The path has crumbled, but there are spectacular views to sea. For some hours, the only people we meet on the windy beach are dog walkers, accompanied by elderly greyhounds. Some have three or four leashed up, one woman has five skittering from side to side.

In one of the deep gaps in the dunes, we talk to a man with a smouldering roll-up and tattered jersey.

"They're mostly retired dogs up from Yarmouth," he says. "There's still one owd boy up the rud who has about fifteen and races 'em," he points. Then in proper Norfolk understatement adds, "Bit blowy, innit?"

We nod. The easterly gale is whipping sand that scours skin, reddens cheeks, and dries eyes.

South of Walcott the low mud cliffs are washed smooth at the base. They are not as majestic as Poppyland's but prone to slip and slide away just the same. A line of protective angled revetments stretches into the distance. They seem to be effective, but then as we approach Happisburgh we see increasing signs of sea damage: a section has not been repaired. Just beyond the village is the mile-long stretch that has completely disappeared, and behind the great curve of cliff eaten away. From the beach, we can see houses hanging over the orange cliffs. At this moment, the tide is low but the waves are being driven by an onshore wind, thumping onto the beach, white froth immediately whipped off before the next wave rears up and crashes onto the sand.

We pass a line of huts on the cliff top called Little Light, where there was once a small lighthouse. Down amongst the surf are concrete and brick remains. Somewhere under the sea lies the ruins of Whimpwell Green church and fishing village, and a little further on those of Eccles church. Visitors once flocked to see its lonely tower on the beach until that too was washed away in 1895. The sandhills moved inland in the 1830s, swamped the church, and then had left it marooned on the beach fifty years later. We stand on the dunes and count seven inland church towers amongst the low woods and broads.

Bush Estate is almost all that remains of Eccles. The fishing village was swept away by a violent storm in 1604. Behind the spectacular dunes and marram grass is a community of extended huts and bungalows, a couple of flags of St George snapping at poles, white vans, and tracks riddled with potholes. How did this settlement come to be built? It is far from villages and schools, but a perfect place for playing on the beach and walking greyhounds. Yet there's no view of the sea from the houses, just the growl of waves from the beach. We walk along the dune tops, with views to the right across the lowlands, and to the left the sands and angry breakers. It is hard walking both on the soft sand and on the tops where the wind is strong. Later we will have smarting eyes, dried as if they had been in a desert rather than by the English seaside on a winter's day. We jump down onto a track behind the dunes to Sea Palling, snobbishly called Appalling-on-sea by some. But here are charming wooden huts on pocket plots, yellow, green and bright blue with outside toilets and felt roofs, named Ravendene, East Dene, and Shangri-la.

The pub is a sprawling mix of wood, glass and gaudy Christmas decorations. There's tape on the carpets, and on the wall photos of the 1953 floods and the lifeboat being launched by horses from the beach. Beside them a red-and-white Santa crawls up the wall by the torn pool table. Sea Palling's lifeboat is special, for in 1930 the RNLI closed the station, only for the local community to re-establish the Palling Voluntary Reserve Service in 1972. They have an inshore boat operated from up on the dunes, 18 volunteers, and are entirely financed by local people. They look out on a sea that hides Haisbro' Sands, a strip of treacherous shifting shallows and deeps extending 9 miles offshore, the safest passage ironically lying along a reach

Happisburgh

called the Devil's Throat. In 1801, *HMS Invincible* foundered here with the loss of 400 lives; 119 of whom were buried in Happisburgh churchyard. We look out of the windows from the otherwise empty pub, and listen to the buffeting wind.

This is holiday land without the heat of summer, but the light today is breathtaking. Across the marshes, the sun seems to skulk behind low ranks of cumulus, and shines a fan of rays onto the landscape of churches, woods of stunted trees and the reed-fringed Broads. Ahead is Waxham, another settlement sheared off by the sea. Its church was swept away in the 1500s, and most of the rest of the village went in a storm of 1791. We pass a sprawling pig farm and walk along a straight road by sheep feeding on beet tops, cars flashing by until one comes to a stop and an elderly couple wind down a window.

"Which way is Sheringham?" they ask innocently.

It's 25 miles away and they have been hoping to come upon it by chance. We point up the road and show them our map, and walk on to Waxham's flint hall and great medieval tithe barn, and then climb into the wild sandhills again. There's now no settlement between here and Winterton-on-Sea. To the right lies Horsey Mere,

Wentworth Day's marshland territory, and the Hundred Stream.

Two people appear out of the dunes and ask, "Have you found them?"

We raise our eyebrows.

"The seals," they say.

There is a colony over towards Winterton Dunes. We point them in the general direction, and then later cut in at Horsey Gap to walk the path behind the hills until we too come to the place. It's a rookery of grey seals, their breeding place. The temperature has slipped a couple of degrees, and the wind bites. Here are pups and adults hauled up on the beach, and some sheltering up here in the dunes. There's a mix of dark browns, greys, blotchy blacks, great 300 kilogram bulls and tiny new pups all in white. We stand close to one older pup as it lies on its side and apparently smiles. Then we see a young pup pushing a piece of black plastic up the beach with its face, hiding behind it to protect its eyes from the sand blasted by the wind. Eye protectors. It looks up, then pops down and pushes, then peers up again to check on progress, and pushes again. And looks up, with that wry upward twist of mouth. How did it learn to do that? None of the other animals takes any notice. An adult female watches us move past, lifts its head, and gazes intently. Yes, we think. What do you want to ask? A large bull splashes into the sea and turns to look, head tipped to one side. Perhaps we are in the zoo.

We leave the seals and head for the final few miles across Winterton Dunes nature reserve to the fishermen's houses of the village itself. There's no sign of the naturists' colony in these old dune warrens. But these embryonic dunes are better known for the natural wildness. They are mainly calcareous with some acid patches, and are home to 170 species of birds. Wentworth Day writes of there being seven or eight sea eagles sailing over these sandhills in the nineteenth century. Perhaps they will return one day. In the humid slacks and shallow ponds are rare natterjack toads, which don't so much hop as run, and so were much sought after in the past: catch a running toad, put it in an ants' nest until the flesh is eaten away, throw the bones in a river at midnight, and then gather the one that floats upstream. The resulting toad bone was strong magic.

An old decoy pond is invaded by low silver birches and twisted oaks. There are fox holes in the surrounding earthworks and thick lichen banks underfoot between buttons of old heather. A marsh harrier climbs from the reeds and then faces motionless into the wind. Down it goes like a black arrow, and then it flips up again. Earlier, we saw an entire flock of lapwings rise from a field and remain still as they turned into the wind. At a certain threshold close to town, dogs begin to appear. Most are not greyhounds. "He won't bite," calls one woman. But then this time the dog scampers by. Winterton is a mix of old and new, and feels like it is at the end of the road. Half-friendly, some smiles, then blank-faced tradesmen in white vans. An elderly woman with vivid purple hair walks up accompanied by panting greyhound and she ties it up outside the shop. We wait and watch life pass by slowly in this edge land.

The end is near. My brother and I are off to California for a descent into memory. All the roads have double-yellows, and we are lucky to find a small municipal park on a cliff top. California has few year-round residents and acres of caravan parks. It was established in 1848 when a cache of sixteenth century gold coins was discovered on the beach. Today the sea is quiet, grey, flat calm it seems, yet the thirty wind turbines over on Scroby Sands are turning slowly. Once those sands remained uncovered by the tides, grass grew and terns nested there. Now they are home to towering turbines that will remain on the horizon for today's walk. We stride along the beach, sometimes on shingle but mostly on the hard sand left by the retreating tide. A row of Norwegian granite boulders protects these cliffs, and we stop to talk to a bowed woman with a hobbling grey-whiskered Norfolk terrier.

"Oh, more than fifteen years," she says with a hint of a Dutch accent, when we ask how long the defences have been here. They have worked, it seems, stopping all cliff erosion. Sometimes in winter, though, the sea scrapes away all the sand, taking the beach down to its clay base.

"Oh no," she says, "but I'd love to," in reply to a question about living on the cliff tops. Her home's a couple of miles inland, but the dog's allergic to grass, so they come to the beach for their daily constitutional.

There's a single dark-headed guillemot on the seashore. It jumps into the water as we approach, and then promptly pops back onto the shingle again, looking lost. Caister is best known as a holiday destination and for its castle, but its sprawling sand dunes are this coast's machair. The cliffs have given way and we walk along the springy grassland, nodding to more locals with their dogs and passing the splendour of stands of harebell, yellow rattle, vetches, thistles and aromatic evening primrose. The rest of Caister drifts by, its empty pubs, playgrounds and site for Roman shore fort, and in caravans lined up to face the water families gather for breakfast, gazing out to the slate-grey sea. We stop at the lifeboat station, bright red and blue in the greens and yellows of the dunescape. Caister men never turn back, they say, and this is another service supported and run by the local community. The first beach company was formed in 1791, evolved into a rescue service, and then the first lifeboat was launched in 1845. But in 1969, when the RNLI decided to close the station, Jack Skipper formed the Caister Volunteer Rescue Service, and its men and women have rescued more than 1,800 people from these waters.

To the south the coastline changes too, no longer sheared off but swelling out into hundreds of hectares of denes that accrete as a result of longshore drift. The roads run north to south, and lead into Yarmouth's 5 mile urban strip bounded half a mile inland by two rivers and the expanse of Breydon Water. We pass a golf course dotted with islands of gorse, Yarmouth race track, and then more caravans. Some have decking, decorations and names whilst others are anonymous. By Newtown, new back in the 1930s with its suburban style and poets' road names, we find a pink-and-purple café. It's busy, a place of choice for local workmen, always a good sign, and we queue for bacon rolls. We study the map, and then we're off to sample Yarmouth's cascade of resort styles. First the line of interlinked pools for pedalos, 1930s-style grassland edged with chunks of pebbled concrete and beds of orange and yellow marigolds. It's clipped, manicured and sparse; an old man stumbles with an arthritic dog. There are hotels and boarding houses, deckchairs on the front, bowling greens and car parks, and people sitting and watching the morning's slow advance.

Brash Britannia pier brings another change, and marks the northern end of the Golden Mile. It is advertising its owner in big letters and bright colours. Yarmouth was the most bombed coastal town in the war, with 8,000 bombs dropped and 2,000 homes destroyed. Come friendly bombs, said John Betjeman about Slough's post-war developments, and the pedestrianised high street here could do with a few. We find ourselves moving swiftly through and into one of the tiny alleys that once defined Yarmouth. At the bridge opposite the Star Hotel, we look downstream along the 3 miles of the Yare that make up the harbour. Upstream at the confluence of the Yare and Bure Rivers stood the Bowling Green pub, once the centre of wildfowling, eeling, smelting and shrimping operations so fondly described by Arthur Patterson. To the south, a solitary fishing boat is moored, and lining the quays are towering red-and-blue tugs and rusty barges, all lifelines to the oil platforms at sea. Yarmouth's prosperity grew with the herring, and they were fished out. It received a boost when North Sea oil was discovered in the mid-1960s, but the town is now having to think about what will happen when oil goes the way of the herring.

A hundred years ago this harbour, like those at Lowestoft, Southwold and Harwich, would have been choked with drifters, smacks and inshore trawlers. In 1913, 820,000 cran of herring were landed here, each cran about a fifth of a ton. Nearly a billion fish landed in one year at one port. A whole culture of practice and language grew up around those wonderfully abundant herren'. Lightly smoked herrings were silver in colour, while those cured for longer turned red. They were smoked using only the shavings and the dust of well-cured oak, which smouldered rather than burned. A single smokey would require 10,000 fish to be put on sticks, a laborious task called speeting the herren'. There was a whole language for counting too. Four herrings to a warp, 6 score to a 100 (actually a long 100 of 120), ten 100s to a cran and 10 crans to the last (roughly two tons in weight). Later, the long 100 had to be changed to 132 fish when the average size of the herring diminished.

Scottish girls came south to gut and speet the herring. They wore balmskins (oilskins), shawls and long skirts, and wrapped their fingers with rags for protection from their fiendishly sharp knives. They smelled so strongly of fish that before the season boarding-house owners stripped their rooms of all furniture. Lowestoft and Yarmouth shopkeepers did well, as the girls regularly sent back goods to their families in Scotland. Many came and some stayed. One, Margaret Rhind, married my grandfather's grandfather in 1840, mariner and train driver. There was also always conflict between the Yarmouth and Lowestoft fisher boys. The Yarmouth boys were said to wear blue jumpers, whilst those of the Lowestoft boys were tan. Nets were treated with cutch, derived from *Acacia catechu*, and the Lowestoft tradition was to dip their jerseys in the cutch to turn them a tan colour. Patterson records, however, wildfowlers and wherrymen of Breydon wearing both blue and tan jerseys, so perhaps group inclusion and exclusion was not so simple. But Yarmouth and Lowestoft have always been in competition. If you are for one, you are against the other. Some of the herring from these ports was exported to the Azores, to southern Spain in the South Spainers, and to the West Indies, the boats then bringing back fruit as fast as they could to prevent decay setting in.

Amid today's warehouses and abandoned concrete quaysides of Southtown we see occasional houses, some stylish for merchants or ship owners, and others tiny cottages for the fishermen and Scottish girls. The brick Royal Artillery building with glass roof above an old drill hall is now converted and advertising careers in

Hopton shore

construction. We climb to the Old Ferry Boat Inn, with its black-and-white mural of a ferryman rowing a small boat: there's no ferry today, though, nor has there been for forty years. The inn and the seaward side of Yarmouth are separated by miles of road but only a couple of hundred metres of water. At the waterfront we pass tug-boat crews on painting duties, and then two men unloading a lorry onto an empty hard covered in weeds. They stop, and look at us guiltily, then aggressively. We switch eyes ahead and keep moving, past the lifeboat, and then finally to the harbour outfall and the northern edge of Gorleston. And within a few metres, we cross a landscape watershed, and ahead are beach and cliff 7 miles south to Lowestoft and its white wind turbine. Before us now is genteel seaside architecture, a wide promenade, grassy slopes, and an elliptic boating pond that holds so much sky. And now comes the rain that has threatened for several hours. An inky cloud pours over the cliff, and the squall has families on the beach scampering for cover. We sit beneath a concrete shelter looking to sea. Before us is an empty blue paddling pool edged with red railings, beyond the ochres of the sand and greys of the sea and sky. The rain passes and as we walk the promenade grades into dunes and cliffs and we've left behind 5 miles of built-up seafront, and a sense of the wild returns.

Above is a white smear of holiday camps on top of the rolling cliffs of Hopton. We walk on the hard sand, and listen to the shingle in the waves. Grey clouds still hang low, spilling rain as they race past. The beach gradually narrows and disappears before Corton, and is replaced by a line of revetments. The red signs say *No way through*, but we press on. The defences have slowed the action of the sea, leaving a no-man's-land between them and the yellow cliffs. The waves crump to the left, the cliffs crumble to the right, clay is exposed beneath our feet. Sand martins flit around us, brought low by the rain. Corton once had a seaward village called Newton, another settlement of this coast under the sea. Here we divert inland, and climb up to farmland fringed by poppies and ox-eye daisies and fields of shimmering barley and yellow-green wheat. Corton's St Bartholomew's Church is half-hidden in the trees, its eyeless and abandoned tower staring out across the cliffs and sea. There are skylarks singing over green hedgerows and yellow fields, under a painter's blue sky now with racing white clouds. Martins chitter over the corn. The glorious Suffolk landscape spreads out before us, and in a tiny back lane we come upon a shrine with three young trees, a bay, pine and spruce, and bright beds full of carefully tended begonias, aubretia, asters and miniature roses. A honeysuckle climbs the hedgerow; a sign says *Welcome to my garden. To Clive*, says another. A sentimental poem is pinned to a board. What happened here, we wonder? A life was lost, and the memorial makers have put energy and emotion into this place. We dart across the busy dual carriageway, and then find another accident site, where a mature oak on a bend is scorched black up one side, and melted debris of car pieces is scattered along the hedgerow.

We walk down a long lane to Blundeston and into clotted memory. The land closes in and it feels like we're walking on the bottom of the sea. Here are houses not seen for thirty years, a forgotten pine plantation, over there the pits we played in, and finally we turn left to find our old road. The farm pond at the end is still here, overhung by an old oak and plane. Golden carp drift in the water. For a moment, I have images of children and leaves and unidentified figures in this glittering light. We walked out there one winter when it froze solid. Things have changed: a new estate and infilled single houses by the farm. We stand across from our house, not wishing to intrude, absorbing every detail. The owner is under the car port at the side. We hesitate, then walk across.

"You're John and Susan's kids, aren't you?" he asks.

They're John and Susan too, and bought the house from our folks in the eighties. It's forty years ago this month since we moved here, and there have only been two owners. We're invited around the back and look at the extension and garden. Jumbled memories crowd in: playing cricket in the garden, the aviary of songbirds in the conservatory, our bedrooms above, building a canoe in the shed, the vegetable garden, the spot where I chased Chris around the house; he ran under some scaffolding and I took it square on my forehead. Snow in the garden, sun in the garden. The memories are leaking out of the brickwork and rising out of the soil. We can't stay too long. We'd drown.

We head for the pub we never went to. Too close to home. It's the Plough with sign stating that this was where Barkis the carrier started in *David Copperfield.* We sit by the stuffed fox on the windowsill. The barman locks the door. Closing time, he says. A Millennium Green has been carved from a field, but near by the children's play area is still as we remember it. Surely that roundabout is far too dangerous for today's parents? We don't walk up to the prison, but we do pass the place where one dark night a policeman leapt out of the hedge at me as I cycled past. Don't worry, he warned, there's a prisoner escaped from up the road. I can remember thinking this is mad, it would be better not to know. We pass the old doctor's hall now refurbished in modern ranchero style with sweeping brick walks and metal gates. The moat, once so carefully maintained, is overgrown. Up the hill we cycled so often, and on the right is the old strawberry farm, and I can still feel the back ache. Chris remembers this is where his hay fever started, probably the mould and dust on the straw. Over to the north was the bulb farm where we laboured for more slim wage packets.

Each step back to the sea is animated by surprises. Up at the crossroads, though, is an indelible memory: a thumping crashing moment when a car piled into my offside, smashing the strut by the driver's door, flipping the white Imp through the air and into a field of golden stubble. I stand for a picture on the roundabout that did eventually replace those near-deadly crossroads. We walk back into Corton and know we're returning to the 1970s. American football was brought to north Suffolk by the oilmen at Yarmouth who wanted a team for their kids so they could play against the four US airbases in the region. But they needed locals to boost the team, and we both played. Those invaders brought a sense of freedom that we'd never experienced. They also brought real hamburgers. In the days before fast food, their great barbecues were a rare novelty. We can still smell the scorched meat and taste the then exotic root beer. Out in the woods today there are kids who'd never believe it. We can hear their shouts and screams from the new theme park. Into Gunton Woods nature reserve; we tramp along twisting paths and past a secluded pond, and then get lost in Gunton Warren, where the bracken stands 8 feet tall on the sandhills. We finally pop out from secret paths on these cliff tops and into north Lowestoft with its wide boulevard, houses with more windows than walls, and arrive at our old school.

And that was it. A year of walking the coast was over.

But I looked at my notes, and realised I hadn't walked to the ends of all the piers to look back towards the coast, as I'd intended, nor gone to Kent to do the same. I also hadn't been out to the forts in the North Sea, up any castles, skidded across the waves in a yacht, or been fishing. I hadn't talked to vineyard owners, or archaeologists about the Neolithic barrows or Roman remains. I'd loved to have met lighthouse keepers but none remain. I hadn't got to the region's last working decoy, or investigated the traditions of folk music. There were pieces of coast missing, Saxon Sutton Hoo, the site of Seahenge. There was the Army garrison too, and those trying to build more housing, and those opposed. I'd missed reserves and habitats, left species unobserved and many wild stories untold. I didn't sight a whale or porpoise, and never found any amber. I checked in my box, and picked up a tiny glass bottle, a head of teasel, an egret's white feather, the dried tangle of seaweed, an ivory Victorian toothbrush, a chunk of bog oak, and a red stone with a hole, all elements of this shore. I still felt the light of this luminous coast was in my eyes and the salt on my skin. I imagined I could hear the mournful music of waders too, and the crunch of shingle on the beach. There seemed much still to discover. This coast might have become a calling.

Chafford Gorges

Coda

I started by a chalk cliff ringed with a patch of scrubby ash trees, ryegrass and mauve chicory sandwiched between the road at my feet and the great orbital above me. Summer had not yet advanced into autumn, but all the horse chestnuts in the region had burned off under leaf-miner attack. I walked up a track behind a chalk cliff that led to the top, and watched the northbound traffic pouring out of the tunnel outfall. But the bridge itself was silent, the traffic becalmed and stretching back into the far distance. Drivers who had seen this before stared blankly, others despaired. I found an apple tree laden with fruit right there by the motorway edge, and then walked east on this ashen morning past new office buildings that had sprung up on some of the many development sites. *Pallets wanted*, said signs on wasteland from which a column of smoke spiralled. *High-profile development*, stated another, *would suit office or hotel.* I could see the red of the detergent factory in the distance, and made for St Clement's Church. I jumped as a high-pitched whistle sounded, and the Paris train flashed towards the new tunnel under the Thames. Lorries were queuing to offload containers onto a temporary storage site by the river, a sheen of hydrocarbons on the polluted soil. But there was no one else on foot. I felt I was a ghost as I walked.

St Clement's was the touch of sanctity amidst the industrial landscape. I walked around the flinty church dwarfed by the red-and-grey factory, on the roof of which a man in fluorescent jacket leant over and fiddled with a panel. The church was tiny; the clanking factory and its neighbouring blue hangar vast. I left the churchyard to the rabbits and walked to the Thames, picking blackberries from the bushes. The sun had barely shone this summer; the berries tasted of dust. On the river wall two fishermen sat under green umbrellas waiting for the tide to cover the mud. I counted seven supermarket trolleys sinking slowly. One of the fishermen had straggly grey hair and a white beard. I tried by the jetty, he said, pointing west, but nothing. Not a bite in half an hour. I nodded, and put my hand to my face, and felt a beard too.

This was strange. I didn't recall ever having had a beard.

I decided to follow his pointing finger and walked west first, past the West Thurrock marshes to Stone Ness where the river turns back north-west towards the bridge and Purfleet. The Thames was at first deserted, then a tug boat surged upriver, propellers thudding, going backwards. I blinked, and looked again. It really was stern first, and a couple of minutes later its wake slapped on the shore near my feet. And then a discovery. In front of the marshes, officially a brownfield site lined up for more development, but also containing a rich mix of insects, including the distinguished jumper spider and hump backed red ant, was an art gallery. A mile and a half of 6-foot river-wall panels had been painted by graffiti artists in every colour, with verve, skill and some poignancy. Most were labelled with dates, 2008 and some 2007, so they were obviously recent. The tags said *Brave One, Vino, Didi.* And there were many mentions of Ozone and Wants: *I miss you; Ozone RIP.*

At a scarlet panel with white letters, *Want-Ozone,* a man with two bull terriers popped his head over the wall. We talked. Twenty-five years ago in East London he used to do this too, but now comes to these marshes to exercise his dogs, chase some rabbits. Geoff said he knew these two lads: Bradley Chapman of Grays, Ozone, was a friend of Banksy, and was killed with Daniel Edgar of Southend by a District line train on New Year's Eve. Geoff said he'd seen the council and river police try to chase off other artists. It sounded like something from an Ealing comedy; he agreed. People came from far and wide: from Ipswich, Colchester, west London, Devon even. Someone had scrawled on top of one of Brave One's pieces, and there was an arrow in purple with an added warning: *Get caught doin this, your goin in the river!!* Despite this, many of the best works were tagged with scribbles. The great dogs rolled at my feet amongst the sea asters and sea blite. But Geoff said if he ever caught someone bombing one of those boys' memorials, he'd set the dogs on them.

The contrast between the art gallery on the shore and the housing on the cliffs at Chafford Gorges could not be greater. I walked up to find another invented community. Eight thousand homes set around a series of linked chalk gorges in what is called Chafford Hundred, the name for one of the original Saxon Hundreds of the county. For a 150 years, the land here was quarried for brick-earth, gravel, flint and chalk. The cement companies bequeathed their names to Wouldham Cliffs and Lion Gorge. Now the Wildlife Trust manages the 200 acres of orchid meadows, willow and oak woodlands, and lakes full of fish and newts. Around the top of Warren Gorge houses crowd to the very edge. They were probably the only homes in the whole of East Anglia perched safely on cliff tops. The warden Adam explained that this was an oddly mobile community. The reserve should be an opportunity to connect local people with nature, but his work was harder because people kept moving. I walked down into the gorge, but felt wobbly and sick. I sat down on a bench. Something about the time felt odd. If only I'd had my American watch, I'd have been able to know the time.

I diverted to Hangman's Wood before heading back into Grays. I wanted to see the famous deneholes set in a mix of urban parkland and unkempt patches of oak and ash. In one glade was a purple tent. A pale woman stood to pull back her hair tight, and a thin man crouched and gazed into the ashes of a fire. In another, a grandmother and a small boy kicked a football back and forth. The diggings originated in Neolithic times, and the chalk pits are full of marks, graffiti and historical clues. At the bottom, they are trefoil shaped, and the bones of animals and humans have been found. The Victorians loved the mystery. Were these deneholes flint or chalk mines, pitfall traps, granaries, refuges or burial chambers? There were more recent deaths here too, and the creak of the hangman's craft seemed to float in the massy air. There were more than seventy shafts in all, but most had been filled in leaving no more than shallow depressions. The five or six that remain were isolated by rows of barbaric metal fencing topped with serrated spikes. I rattled a couple of gates, but there was no way in.

I walked back into Grays, the wind by now gusting hard, and turned up Quarry Hill. There was a bend in the road, and when I came round it, here was the building I was looking for. As I walked, slowing down, my feet increasingly sore, the police station seemed to change its appearance. A tangle of blue bars guided visitors up the zig-zagging slope. The glass doors slid open automatically with a clunk. I hesitated and then went inside. The waiting room was empty. Sitting behind the glass panel was a policewoman with chubby cheeks, her eyes hidden behind large glasses. She seemed to be a sergeant, and the name on her white shirt was Norton.

Her expression was unexpectedly reassuring.

"Is it about a bicycle?" she asked.

Notes

PREFACE

page 14 — Trees and fractal surfaces: in Richard Mabey's *Beechcombings* (2007).

page 15 — Listening to other people's ghosts: from Marina Warner's introduction to Memory Maps on V & A website.

page 16 — The Thoreau quote is from his essay "Walking", and Edward Abbey's from *Desert Solitaire* (1968). Ronald Blythe's quote about driving and the depeopled countryside is from his *Going to Meet George* (1991).

page 16 — For a selection on wildness and its capacity to change us: from the UK see Macfarlane's *The Wild Places* (2007), Deakin's *Wildwood* (2007) and *Waterlog* (2000), Griffiths' *Wild* (2006), Cocker's *Crow Country* (2007), Mabey's *Beechcombings* (2007); from North America see Dillard's *Teaching a Stone to Talk* (1982) and *Pilgrim at Tinker Creek* (1990), Lopez's *Arctic Dreams* (1983) and *Crossing Open Ground* (1988), Abbey's *Desert Solitaire* (1968), Finch's *Outlands* (1986), Basso's *Wisdom Sits in Places* (1996), and Nelson's *Make Prayers to the Raven* (1983).

page 17 — Exploring means changing the world: in Robin Hanbury-Tenison's *The Seventy Great Journeys* (2006). For a selection of material on the history of walking, see Marples' *Shanks's Pony* (1959), Taplin's *The English Path* (1979), Solnit's *Wanderlust* (2001), Tait's *Walker's Companion* (2003), Blythe's *Field Work* (2007), Nicholson's *Lost Art of Walking* (2008), and especially the section on walking in Chatwin's *Songlines* (1998). Stevenson and Hazlitt wrote essays on the values and experiences of walking.

page 17 — Reinhabitation is a term of Gary Snyder's: see *The Practice of the Wild* (1990) and *A Place in Space* (1995).

page 18 — These stones and things talk: see Annie Dillard's *Teaching a Stone to Talk* (1982). Common Ground celebrates mementos of place in the Real Souvenirs campaign: "A pebble is just a piece of stone, but it is transformed because of the aura of memory."

CHAPTER 1

page 23 — For more on the consequences of modern consumption patterns, see Pretty (2007), Diamond (2005), Lovelock (2005), McKibben (2003), Rees (2003), McNeill (2002), Davies (2002) and Jacob (1961). On the psychological drivers, see Frank (1999), Kasser (2002), Nettle (2005), and on the economic and political, see Gray (2002, 2004), Orr (2004) and McKibben (2007).

page 24 — See Iain Sinclair's *London Orbital* (2002).

page 24 — The River Thames: one of the dark places of the earth: from Joseph Conrad's *Heart of Darkness* (1902). Accounts of Thameside can be found in Morgan's *Forgotten Thameside* (1951), Wright's *The River* (1999), Ackroyd's *Thames* (2007), Maxwell's *Unknown Essex* (1925) and Barnes' *Grays Thurrock Revisited* (1991). Sinclair walked Grays part of *London Orbital* (2002).

page 27 — The Tilbury foxes are in Wentworth Day's *They Walk the Wild Places* (1956).

page 29 — The gull's clanging cry: in Crabbe's *The Borough*, where the cry is from a goldeneye.

page 30 — Death and placeless world: see Robert Pogue Harrison's *Dominion of the Dead* (2003). Also see Worpole's *Lost Landscapes* (2003), and Clark and Franzman (2009) on roadside memorials.

page 33 — John Tusa on Bataville: from University of Essex Burrows Lecture (2003) "Shoemakers to the World". At www.essex.ac.uk/burrows/. See also Wright's *The River* (1999), and the Bata Reminiscence and Resource Centre at http://www.batamemories.org.uk/. Vikram Seth was a Bata child in India.

page 34 — The Shell Haven and Kynochtown history is explored in Barnes (1991). See also Morgan's *Forgotten Thameside* (1953).

p.37-38 — For Canvey's original guide, see Basil Cracknell's *Canvey Island* (1910), Lucy Harrison's *Captivating Canvey* (2007).

The Coulson Kernahan quote from 1897 appears in the Lucy and Gould *Anthology of Essex* (1911). Canvey on a moon-misty night: see Wentworth Day's *Coastal Adventure* (1949).

CHAPTER 2

page 43 — Peter Ackroyd's *Thames* (2007) describes the establishment and use of prison hulks on this part of the Thames by Canvey, and the use of forced labour and chain gangs.

page 47 — Many have written of the effects of the 1953 floods: there are illuminating and revealing reports to be found on the web. But Hilda Grieve's *The Great Tide* (1959) is an outstanding account of the 1953 floods on the county of Essex. Photographs can be found in Currie et al. (1992), and further accounts in Jarvis (1992). The story of the *Princess Victoria* disaster is in Hunter (1999) and Cameron (2002).

CHAPTER 3

page 61 — A general account of the emergence of the seaside as a cultural phenomenon appears in Lencek and Bosker's *The Beach* (1998). For Southend holidays, see Everitt (1980) and for the story of the Belle steamers see Box (1983). For a general history of the coast, see Cracknell (2005), Gale (2000); for the regionally specific, see Dutt (1910) and Leather (1979). For an offshore view of the coast, see Raban's *Coasting* (1986), by rail, foot and bus, Thoreau's *Kingdom by the Sea* (1983).

page 64 — On the plotlands of Basildon, see Walker (2001).

page 66 — Children and outdoor play: see Richard Louv's *Last Child in the Woods* (2005), and Nabhan and Trimble's *Geography of Childhood* (1994). Also see Pretty *The Earth Only Endures* (2007).

page 68 — Billy Bragg and the A13 (2006): an essay from the V & A Memory Maps website.

CHAPTER 4

page 81 — Our lives, our tiny lives is from Annie Dillard's *Teaching a Stone to Talk* (1982). The fishermen at Pakefield: from W. G. Sebald's *Rings of Saturn* (2002). On fish kettles, see Benham (1971).

page 83 — For more on life on the Dengie, see Bruce (1971), on Essex food, Pewsey (1977), and on the coast, Worpole (2006) and Orton and Worpole (2005). On bees, see Benton (2006). Samuel Bensusan moved to Bradwell on the Dengie in the 1880s, and write some 500 stories about the local life: see his *Village Idylls* (1926), *Back of Beyond* (1945), *Salt of the Marshes* (1949), *Marshland Omnibus* (1954) and *Marshland Voices* (1955). The oxbird pudden is described in Wentworth Day's *Book of Essex* (1979). Slavery of farm workers: in S. L. Bensusan's *Back of Beyond* (1945).

page 84 — Samphire redolent of iodine: in Richard Mabey's *Flora Britannica* (1996). John Norden's quote is from Ronald Blythe's *In Praise of Essex* (1998). J. A. Baker's book of the dead is his *The Peregrine* (2005). Robert Macfarlane wrote the introduction for the latest edition. The Bensusan quote about the "ellum" is from *Back of Beyond* (1945).

page 86 — Wildfowling on the Blackwater: see Wentworth Day's *Marshland Adventure* (1950), *Coastal Adventure* (1949), and *Rum Own Boys* (1974). See also Leather (1979). An excellent history of duck decoys appears in Heaton (2001).

page 89 — "Yet this will go onward the same, though dynasties pass": from *In Time of "The Breaking of Nations"* by Thomas Hardy.

page 90 — For a history of the Peculiar People, see Sorrell (1979), and for the Othona Community, see Motley (1985).

page 91-92 — On views across estuaries: see S. L. Bensusan's *Back of Beyond* (1949).

page 94 — The field of stag-horned oaks of Mundon: see Wentworth Day's *Coastal Adventure* (1949) and Robert Macfarlane's *The Wild Places* (2007).

page 97 — Stuck fast in the muds: from Wentworth-Day's *Coastal Adventure* (1949).

CHAPTER 5

page 103 — The land experienced as terrain: from Worpole (2006) and Orton and Worpole (2005).

page 104 — The bleak and treeless Essex archipelago : see Wentworth Day's *They Walk the Wild Places* (1956).

page 107 — Smuggling in Essex: see Smith (2005). On witches and devils: see Wentworth Day's *Book of Essex* (1979) and *Marshland Adventure* (1950). The wonderful account of the duck-hunting foray into Kent appears as the final chapter in *Marshland Adventure.* For Cunning Murrell, see Morrison (1900).

page 111 — For a millennial history of Tollesbury, see Crossley, Dunn and Heard (2000). Leather's *Salty Shore* (1979) is about the Blackwater Estuary.

page 114 — Ague in every family: from Herbert Tompkins' *Marsh Country Rambles* (1904). Defoe's account of all those dying Essex wives is in his *Tour Through the Whole Island of Great Britain* (1724-26). The 1594 John Norden quote on the ague is from Lucy and Gould's *Anthology of Essex* (1911).

page 116 — See Smith (1970) for a history of Foulness Island.

CHAPTER 6

page 121 — On the performance of nightingales: see Mark Cocker and Richard Mabey's *Birds Britannica* (2005), and Mabey's *Book of Nightingales* (1997).

page 122 — Wentworth Day sighting of sea eagles over Geedon is reported in *Farming Adventure* (1941).

page 126 — Herbert Tompkins came to Colchester for his *Marsh Country Rambles* (1904).

page 129 — On oysters in the Colne: see Bolitho's *The Glorious Oyster* (1929) and Younge's *Oysters* (1960). Also Leather's *Salty Shore* (1979), and Roberts and Roberts' *Paglesham Natives* (2006).

page 131 — Pistol mentions oysters to Falstaff in *The Merry Wives of Windsor* by William Shakespeare (1600): Falstaff: "I will not lend thee a penny." Pistol: "Why, then, the world's mine oyster, which I with sword will open." Falstaff: "Not a penny."

CHAPTER 7

page 135 — Roger Deakin made a 2005 BBC Radio 4 programme about his literary canoe journey in his boat called *Cigarette.*

page 136 — On Stour Valley history: see Waller (1959), Benham (1971), Salmon (1977), Ronald Blythe's *Word from Wormingford* series, and also *Field Work* (2007) and *A Year at Bottengoms Farm* (2006). See also Mee (1941) and Scarfe (1960).

page 136 — The story about the dragon of Bures and Wormingford appears in Blythe's *In Praise of Essex* (1988).

page 138 — Red squirrels in the Stour: see Waller (1959). The Constable quote about these scenes making him a painter is also from Waller.

page 139 — The stoat doesn't always win. Thomas Coward's "Rabbit versus Stoats" appeared in July 1905 (in Wainwright, 2006), and he tells how a single rabbit attacked a family of stoats, and saw them all off. Other accounts of stoats dancing and hunting are also by Arthur Boyd (1939) and William Condray (1973).

page 139-140 — The meals at Tendring Hall are described in Waller (1959).

page 143 — See Robert Pogue Harrison's *Dominion of the Dead* (2003) for more on death, identity and place.

page 148 — Running across the muds: in Griff Rhys Jones' autobiography *Semi-Detached* (2006).

page 150 — *Living Like Weasels* is in Annie Dillard's *Teaching a Stone to Talk* (1982).

page 151 — Brown-sailed barges on the Orwell: in William Dutt's *The Norfolk and Suffolk Coast* (1910).

CHAPTER 8

page 156 — A good source for Orford Ness military and natural history is the National Trust's website: http://www.nationaltrust.org.uk/main/w-vh/w-visits/w-findaplace/w-orfordness.htm.

page 159 — For a history of Foulness: see Smith (1970). The Foulness Conservation and Archaeological Society has done much to record and analyse the history of Foulness Island.

page 163 — For a history of East Anglia at war: see Derek Johnson's *East Anglia at War* (1992).

CHAPTER 9

page 169 — The Felixstowe coast: see Ward Lock (1954). The Suffolk coast, see Mee (1941), Dutt (1910)

page 170 — Swinburne's miles of desolation is in Grigson (1980).

page 172 — The Sebald quotes at Shingle Street are from *The Rings of Saturn* (2002).

page 173 — For a history of the Suffolk Punch, see George Ewart Evans' *The Horse and the Furrow* (1960). The Suffolk Punch Trust is at http://www.suffolkpunchtrust.org/. The horse colour is called "chestnut, not chestnut."

page 175 — Griff Rhys Jones' family holidays are described in *Semi-Detached* (2006).

page 176 — Edward Fitzgerald's Aldeburgh sea is in Arnott (1973). Dutt's description of Slaughden is in his *Norfolk and Suffolk Coast* (1910). For a history of the Alde Estuary and River: see Tennyson's *Suffolk Scene* (1939), and Arnott's *Alde Estuary* (1973). Cod boats are described in Harvey Benham's *Once Upon a Tide* (1971).

page 179 — Ted Hughes' wet-footed god of the horizons: the curlew – discussed in Cocker and Mabey's *Birds Britannica* (2005). From the poem "Curlews in April": "Hang their harps over the misty valleys/A wobbling water-call/A wet-footed god of the horizons."

page 180 — Julian Tennyson's *Suffolk Scene* describes wildfowling on the Alde and along the coast. Further north, Arthur Patterson does the same for Breydon Water in *Wildfowlers and Poachers* (1929).

page 182 — Aldeburgh lifeboat station: see http://aldeburghlifeboat.org.uk/.

CHAPTER 10

page 192 — Griff Rhys Jones on loss and memory is from his *Semi-Detached* (2006).

page 192 — For more on the lost villages of England, see Pretty (2007), Driver (2006), Beresford (1954) and Beresford and Hurst (1979).

page 193 — Swinburne on Dunwich: see Grigson (1980). Henry James on Dunwich: "I defy anyone …" appeared in his essay "Old Suffolk" in his book *English Hours* (1905).

page 195 — For a history of Southwold railway: see Taylor and Tonks (1979), and for Southwold, see Clegg and Clegg (1999). See also Sebald's *Rings of Saturn* (2002).

page 203 — William Dutt on Lowestoft's fishing community – see Dutt (1904) and (1910).

CHAPTER 11

page 207 — See Robert Macfarlane (2008) for an account of traditional Norfolk farmers, which includes photographs by Justin Partyka.

page 210 — For a history of the Norfolk coast: see Storey *The Lost Coast of Norfolk* (2006), and Dutt (1910).

page 213 — For more on Abbey Farm organics: see http://www.abbeyfarm.co.uk/visiting.shtml; and on pinks, see Cocker and Mabey's *Birds Britannica* (2005). The Aldo Leopold quotes are from essays in Leopold's *Sand County Almanac* (1949). Gallico's *Snow Goose* was published in 1941.

CHAPTER 12

page 229 — An account of Poppyland appears in Blythe's *Going to Meet George* (1991).

page 232 — Martin Newell's poem "Black Shuck" (1999) appears in a book with atmospheric linocuts by James Dodds.

page 234 — The gasworks are in Ewan MacColl's song "Dirty Old Town" (1949).

page 234 — For more on Happisburgh: see Coastal Concern Action Group at http://www.happisburgh.org.uk/.

page 236 — See Ronald Blythe's excellent essay on P.H. Emerson in *Field Work* (2007). For more on reed-cutters in Norfolk: see Broads Reed & Sedge Cutters Association at http://www.reedcutters.norfolkbroads.com/ and North Norfolk Reed-cutters Association at http://www.norfolkreed.co.uk/. A compelling account of Hickling, Horsey and other Broads can be found in Wentworth Day's *Marshland Adventure* (1950). See Dutt (1910) for more on the Norfolk coast.

CHAPTER 13

page 251 — For an extended account of Yarmouth and the herren', see George Ewart Evans' *The Days That We Have Seen* (1975), and Dutt (1910), Benham (1971) and Storey (2006).

CHAPTER 14

page 258 — Banksy needs no reference, but Brave One's graffiti art can be seen at http://www.braveone.co.uk/.

page 258 — Chafford Gorges is an Essex Wildlife Trust reserve.

page 259 — The real policewoman in Grays station, Carol, was gracious and helpful, and of course looked nothing like Sergeant Pluck.

Bibliography

Journeys and Pathways

Bailey L. 1964. *The Long Walk.* Westernlore Press, Tuscon

Basho M. 1966. *The Narrow Road to the Deep North and Other Travel Sketches.* Penguin, London

Brody H. 1981. *Maps and Dreams.* Pantheon, New York

Brown D. 1970. *Bury My Heart at Wounded Knee.* Vintage, London

Brown D. 2004. *The American West.* Pocket Books, London

Bryson B. 1997. *A Walk in the Woods.* Doubleday, New York

Chatwin B. 1998. *Songlines.* Picador, London

Cobb D. 2006. Spring journey to the Saxon shore. In *Business in Eden.* Equinox Press, Braintree

Coverley M. 2006. *Psychogeography.* Pocket Essentials, Harpenden

Fiennes R. 2003. *Beyond the Limits.* Little, Brown, London

Finkelstein M. and Stone J. 2004. *Paddling the Boreal Forest.* Natural Heritage Books, Toronto

Hanbury-Tenison R. 2006. *The Seventy Great Journeys.* Thames and Hudson, London

Hazlitt W. 1822. *On Going on a Journey.* New Monthly Magazine, London

Kamler K. 2004. *Surviving the Extremes.* Penguin, London

Leslie E. E. 1988. *Desperate Journeys, Abandoned Souls.* Mariner, Boston

Marples M. 1959. *Shanks's Pony. A Study of Walking.* J. M. Dent, London

McKibben W. 2000. *Long Distance: A Year of Living Strenuously.* Simon and Schuster, New York

Muir J. 1911. *My First Summer in the Sierra.* Houghton Mifflin, Boston (reprinted in 1988 by Canongate Classics, Edinburgh)

Muir J. 1992. *The Eight Wilderness-Discovery Books.* Diaden Books, London and Seattle

Nicholson G. 2008. *The Lost Art of Walking.* Riverhead, London

Santoka Tenda. 2009. *Mountain Tasting.* White Pine Press, Buffalo

Sebald W. G. 2002. *The Rings of Saturn.* Vintage, London
Sinclair I. 2002. *London Orbital.* Penguin, London
Solnit R. 2001. *Wanderlust. A History of Walking.* Verso, London
Solnit R. 2005. *A Field Guide to Getting Lost.* Canongate, London
Stevenson R. L.1850-94. *Walking Tours.* InVirginibus Puerisque, London
Tait M. 2003. *The Walker's Companion.* Think Books, London
Taplin K. 1979. *The English Path.* Boydell Press, Ipswich
Thoreau H. D. 1902. *Walden or Life in the Woods.* Henry Frowde, Oxford University Press, London, New York and Toronto
Thubron C. 2006. *The Silk Road.* Chatto and 'Windus, London
Tompkins H. 1904. *Marsh Country Rambles.* Chatto and Windus, London

Wild and Place

Abbey E. 1968. *Desert Solitaire.* Simon and Schuster, New York
Baker J. A. 1967 [2005]. *The Peregrine.* New York Review Book, New York
Barnes S. 2004. *How to be a Bad Birdwatcher.* Short Books, London
Barron D. 2004. *Beast in the Garden.* W. W. Norton, New York
Bate J. 2001. *The Song of the Earth.* Picador, London
Bergman C. 2003. *Wild Echoes.* University of Illinois Press, Urbana and Chicago
Callicott J. B. and Nelson M. page (eds). 1998. *The Great New Wilderness Debate.* University of Georgia Press, Athens and London
Clayton S and Opotow S. (eds). 2003. *Identity and the Natural Environment.* MIT Press, Boston
Cocker M. and Mabey R. 2005. *Birds Britannica.* Chatto and Windus, London
Cocker M. 2007. *Crow Country.* Jonathan Cape, London
Corbett J. 1946. *Man Eaters of Kumaon.* Oxford University Press Inc, New York
Deakin R. 2000. *Waterlog.* Vintage, London
Deakin R. 2007. *Wildwood: A Journey Through Trees.* Hamish Hamilton, London
Dillard A. 1982. *Teaching a Stone to Talk.* Harper Perennial, New York
Dillard A. 1990. *Pilgrim at Tinker Creek.* (In *Three by Annie Dillard,* 2001). Harper Perennial, New York
Gallico page 1941. *The Snow Goose.* Penguin, London
Griffiths J. 2006. *Wild. An Elemental Journey.* Hamish Hamilton, London
Grigson G. 1988. *Faber Books of Poems and Places.* Faber & Faber, London
Hansen K. 1992. *Cougar: American Lion.* Northland Publ, Flagstaff
Harding S. 2005. *Animate Earth.* Green Books, Devon
Hardy T. 1981. *Landscape Poets* (Introduced by Porter P). Weidenfeld and Nicolson, London
Harrison R.P. 2003. *The Dominion of the Dead.* University of Chicago Press, Chicago and London
Hoagland E. 1972. *Red Wolves and Black Bears.* Lyons and Burford, New York
Hogue L. 2000. *All the Wild and Lonely Places.* Island Press, Washington DC
Jackson W. 1994. *Becoming Native to this Place.* University Press of Kentucky, Lexington
Jones K. 2004. *Wolf Mountains.* University of Calgary Press, Calgary
Lopez B. 1978. *Of Wolves and Men.* Simon and Schuster, New York
Lopez B. and Gwartney D. (eds). 2006. *Home Ground. Language for an American Landscape.* Trinity University Press, San Antonio
Mabey R. 1996. *Flora Britannica.* Sinclair-Stevenson, London
Mabey R. 1997. *The Books of Nightingales.* Sinclair-Stevenson, London
Mabey R. 2007. *Beechcombings.* Chatto and Windus, London
Macdonald H. 2006. *Falcon.* Reaktion Books, London
Macfarlane R. 2003. *Mountains of the Mind.* Granta, London

Macfarlane R. 2007. *The Wild Places.* Granta, London
Nash R. 1973. *Wilderness and the American Mind.* Yale University Press, New Haven CT
Nethersole-Thompson D. 1951. *The Greenshank.* Collins, London
Pretty J. 2007. *The Earth Only Endures. On Reconnecting with Nature and Our Place In It.* Earthscan, London
Rothenberg D. and Prior W. (eds). 2004. *Writing the Future.* MIT Press, Cambridge, MA
Snyder G. 1990. *The Practice of the Wild.* Shoemaker Hoard, Washington DC
Snyder G. 1995. *A Place in Space.* Counterpoint, New York
Thoreau H D. 1837-1853. *The Writings of H. D. Thoreau* Volumes 1-6 (published 1981 to 2000). Princeton University Press, Princeton, NJ
Tuan Y-F. 1977. *Sense and Place.* University of Minnesota Press, Minneapolis
Wainwright M (ed). 2006. *A Gleaming Landscape. A Hundred Years of the Guardian's Country Diary.* Aurum Press, London
Warner M. 2003. *Signs and Wonders.* Vintage, London
Wentworth Day J. 1956. *They Walk the Wild Places.* Blandford Press, London

Nature and Human Connections

Abram D. 1996. *The Spell of the Sensuous.* Vintage Books, New York
Adams W. M. 1996. *Future Nature: A Vision for Conservation.* Earthscan, London
Bass R. 2004. *Caribou Rising.* Sierra Club Books, San Francisco
Basso K. 1996. *Wisdom Sits in Places.* University of New Mexico Press, Albuquerque
Bell M. M. 1997. "The ghosts of place". *Theory and Society* 26, 813-836
Berkes, F. 1999. *Sacred Ecology.* Taylor and Francis, Philadelphia.
Brody H. 2002. *The Other Side of Eden.* Faber & Faber, London
Cooper-Marcus C. and Barnes M. 1999. *Healing Gardens.* John Wiley. New York
Freeman H. 1984. (ed) *Mental Health and the Environment.* Churchill Livingstone, London
Frumkin H. (ed). 2005. *Environmental Health.* Jossey-Bass, San Francisco
Gallagher W. 1994. *The Power of Place.* Harper Perennial, New York
Harding S. 2005. *Animate Earth.* Green Books, Devon
Jackson W. 1994. *Becoming Native to this Place.* University Press of Kentucky, Lexington
Kaplan R. and Kaplan S. 1989. *The Experience of Nature.* Cambridge University Press, Cambridge
Kastner J. and Wallis B. 1998. *Land and Environmental Art.* Phaidon Press, London
Kellert S. R. and Wilson E. O. 1993. (eds) *The Biophilia Hypothesis.* Island Press, Washington DC
Lee R. B. and Daly R. (eds). 1999. *The Cambridge Encyclopedia of Hunters and Gatherers.* Cambridge University Press, Cambridge
Long R. 2002. *Walking the Line.* Thames and Hudson, London
Lopez B. 1986. *Arctic Dreams.* Harvill, London
Lopez B. 1988. *Crossing Open Ground.* Picador, London
Lopez B. 1998. *About This Life.* Harvill, London
Louv R. 2005. *Last Child in the Woods.* Algonquin Press, Chapel Hill
Nabhan G.P. and Trimble S. 1994. *The Geography of Childhood.* Beacon Press, Boston
Nabhan G. 1982. *The Desert Smells Like Rain.* North Point Press, San Francisco, California
Nelson R. 1983. *Make Prayers to the Raven.* University of Chicago Press, Chicago
Orr D. W. 1993. *Ecological Literacy.* State University of New York Press, New York
Orr D. W. 2004. *The Last Refuge.* Island Press, Washington DC

Posey D (ed). 1999. *Cultural and Spiritual Values of Biodiversity.* IT Publications and UNEP, London.
Pretty J. 1995. *Regenerating Agriculture.* Earthscan, London
Pretty J. 1998. *The Living Land.* Earthscan, London
Pretty J. 2002. *Agri-Culture. Reconnecting People, Land and Nature.* Earthscan, London
Pretty J. 2007. *The Earth Only Endures. On Reconnecting with Nature and Our Place In It.* Earthscan, London
Serpell J. 1996. *In the Company of Animals.* Cambridge University Press, Cambridge
Spirn A. W. 1998. *The Language of Landscape.* Yale University Press, New Haven
Stegner W. 1962. *Wolf Willow.* Penguin, London
Suzuki D. 1997. *Sacred Balance.* Bantam Books, London
Thomas K. 1983. *Man and the Natural* World. Penguin, London
Tredinnick M. 2005. *The Land's Wild Music.* Trinity University Press, San Antonio
Wilson E. O. 1984. *Biophilia.* Harvard University Press: Cambridge, MA

Memory, Story and Time

Aries P. 1974. *Western Attitudes Toward Death.* Johns Hopkins University Press, Baltimore
Bergsten S. 1973. *Time and Eternity.* Humanities Press, New York
Blamires H. 1969. *Word Unheard. A Guide through Eliot's Four Quartets.* Methuen, London
Blythe R. 1979. *The View in Winter.* Canterbury Press [2005], Norwich
Booker C. 2004. *The Seven Basic Plots.* Continuum, London
Clayton S. and Opotow S. (eds). 2003. *Identity and the Natural Environment.* MIT Press, Boston
Clark J. and Franzman M. 2009. "The making of roadside memorials". In Earle S., Kamaromy C. and Bartholomew C. (eds). *Death and Dying: A Reader.* Sage, London
Harrison R.P. 2003. *The Dominion of the Dead.* University of Chicago Press, Chicago and London
McKee R. 1999. *Story.* Methuen, London
O'Brian F. 1967. *The Third Policeman.* Picador, London
Radstone S. (ed). 2000. *Memory and Methodology.* Berg, Oxford
Radstone S. and Hodgin K. (eds). 2006. *Memory Cultures.* Transaction, New Brunswick
Schama S. 1996. *Landscape and Memory.* Fontana Press, London
Stewart page J. and Strathern A. (eds). 2003. *Landscape, Memory and History.* Pluto Press, London
Warner M. 2006. *What are Memory Maps?* V & A Museum. At URL www.vam.ac.uk/activ_events/adult_resources/memory_maps
Worpole K. 2003. *Last Landscapes.* Reaktion Books, London
Zipes J. 1979. *Breaking the Magic Spell.* University of Kentucky Press, Lexington

Industrialisation and Civilisations

Barlett P. (ed). 2005. *Urban Place.* MIT Press, Cambridge MA
Beresford M. W. and Hurst J. 1979. *Deserted Medieval Villages.* St Martin's Press, New York
Beresford M. W. 1954. *The Lost Villages of England.* Lutterworth Press, London
Berry T. 1999. *The Great Work.* Bell Tower, New York
Davis M. 2002. *Dead Cities.* The New Press, New York
Diamond J. 2005. *Collapse: How Societies Choose to Fail or Survive.* Penguin, London
Frank R. H. 1999. *Luxury Fever.* Princeton University Press, Princeton
Frumkin H., Frank L. and Jackson R. 2004. *Urban Sprawl and Public Health.* MIT Press, Cambridge, MA
Gray J. 2002. *Straw Dogs.* Granta Books, London
Gray J. 2004. *Heresies. Against Progress and Other Illusions.* Granta, London

Halpern D. 1995. *Mental Health and the Built Environment.* Taylor & Francis, Bristol
Haywood J. 2005. *The Penguin Historical Atlas of Ancient Civilisations.* Penguin, London
Jacobs J. 1961. *The Life and Death of Great American Cities.* Random House, London
Kasser T. 2002. *The High Price of Materialism.* MIT Press, Cambridge, MA
Kay J. H. 1997. *Asphalt Nation.* Crown, New York
Lovelock J. 1995. *The Age of Gaia.* Oxford University Press, Oxford
Lovelock J. 2005. *The Revenge of Gaia.* Allen Lane, London
McKibben B. 2003. *The End of Nature.* Bloomsbury, London
McKibben B. 2007. *Deep Economy.* Holt, New York
McNeill J. 2000. *Something New Under the Sun.* Penguin, London
Myers N. and Kent J. 2004. *The New Consumers.* Island Press, Washington DC
Nettle D. 2005. *Happiness.* Oxford University Press, Oxford
Orr D. W. 2006. *Design on the Edge.* MIT Press, Cambridge MA
Quammen D. 2004. *Monster of God.* Hutchinson, London
Rees M. 2003. *Our Final Century.* Arrow Books, London
Rees W., Wackernagel M. and Testemae P. 1996. *Our Ecological Footprint.* New Society Publ, Gabriola
Warner M. 2000. *No Go to the Bogeyman.* Vintage, London

Beach, Coast and River

Ackroyd P. 2007. *Thames. Sacred River.* Chatto and Windus, London
Box P. 1983. *Belles of the East Coast.* Tyndall and Panda Press, Lowestoft
Cameron S. 2002. *Death in the North Channel: The Loss of the Princess Victoria.* Colourpoint Books
Conrad J. 1902 [2007]. *Heart of Darkness.* Penguin Classics, London
Cracknell B. E. 2005. *Outrageous Waves.* Phillimore, Chichester
Driver L. 2006. *The Lost Villages of England.* New Holland, London
Dutt W. A. 1910. *The Norfolk and Suffolk Coast.* Frederick Stokes, New York
Everitt S. 1980. *Southend Seaside Holiday.* Phillimore, Chichester
Finch R. 1986. *Outlands: Journeys to the Outer Edge of Cape Cod.* David Godine, Boston
Gale A. 2000. *Britain's Historic Coast.* Tempus, Stroud
Grieve H. 1959. *The Great Tide.* Essex County Council, Chelmsford
Hay J. 1963. *The Great Beach.* W. W. Norton, New York
Hunter J. 1999. *Loss of the Princess Victoria.* Stranraer and District Local History Trust.
Morgan G. H. 1951 [1968]. *Forgotten Thameside.* Letchworth Press, Letchworth
Leather J. 1979. *The Salty Shore.* Terrence Dalton, Lavenham
Lencek L. and Bosker G. 1998. *The Beach. The History of Paradise on Earth.* Secker and Warburg, London
Ogburn C. 1966. *Winter Beach.* William Morrow, New York
Owens J. G. and Case G. O. 1908. *Coastal Erosion and Foreshore Protection.* St Bride's Press, London
Raban J. 1986. *Coasting.* Picador, London
Thoreau P. 1983. *The Kingdom by the Sea.* Penguin, London
Wright page 1999. *The River: The Thames in Our Time.* BBC, London
Younge C. M. 1960. *Oyster.* Collins, London

Wildfowling and Hunting

Abbey E. 1996. "Blood sport". In Petersen D. (ed). *A Hunter's Heart.* Henry Holt, New York
Animal Studies Group, 2006. *Killing Animals.* University of Illinois Press, Urbana and Chicago
Causey A. 1996. "Is hunting ethical?" In Petersen D. (ed). *A Hunter's Heart.* Henry Holt, New York
Heaton A. 2001. *Duck Decoys.* Shire Books, Bucks
Madson J. 1996. "Why men hunt". In

Petersen D. (ed). *A Hunter's Heart.* Henry Holt, New York
Marks S. 1991. *Southern Hunting in Black and White.* Princeton University Press, Princeton NJ
Nelson R. 1987. *Heart and Blood. Living with Deer.* Vintage, New York
Patterson A. H. 1929. *Wildfowlers and Poachers.* Ashford Press [1988], Southampton
Petersen D (ed). 1996. *A Hunter's Heart.* Henry Holt, New York
Tennyson J. 1938. *Rough Shooting.* Adam and Charles Black, London
Tennyson J. 1939. *Suffolk Scene. A Book of Description and Adventure.* Blackie and Son, London
Wentworth Day J. 1949. *Coastal Adventure.* Harrap, London
Wentworth Day J. 1950. *Marshland Adventure.* Harrap, London
Wentworth Day J. 1974. *Rum Owd Boys.* East Anglian Magazine Ltd, Ipswich
Woods B. 1996. "The hunting problem". In Petersen D. (ed). *A Hunter's Heart.* Henry Holt, New York

Essex, Suffolk and Norfolk

Arnott W. G. 1973. *Alde Estuary.* Boydell Press, Ipswich
Baring-Gould S. 1880. [1998]. *Mehalah.* Praxis, London
Barnes B. 1991. *Grays Thurrock Revisited.* Phillimore, Chichester
Benham H. 1971. *Once Upon a Tide.* Harrap, London
Bensusan S. L. 1926. *Village Idylls.* Noel Douglas, London
Bensusan S. L. 1945. *Back of Beyond.* Blandford Press, London
Bensusan S. L. 1949. *Salt of the Marshes.* Routledge and Kegan Paul, London
Bensusan S. L. 1954. *A Marshland Omnibus.* Duckworth, London
Bensusan S. L. 1955. *Marshland Voices.* Gerald Duckworth, London
Benton T. 2006. *Bumblebees.* Collins, London
Blythe R. 1988. *In Praise of Essex. An Anthology.* Alastair Press, Bury St Edmunds
Blythe R. 1991. *Going to Meet George.* Long Barn Books, Ebrington
Blythe R. 1997. *Word from Wormingford.* Penguin, London
Blythe R. 2005. "Return to Akenfield". In Granta 90: *Country Life.* Granta, London
Blythe R. 2006. *A Year at Bottengoms Farm.* Canterbury Press, Norwich
Blythe R. 2007. *Field Work.* Black Dog Books, Norwich
Boggis H. 2008. *A Country Family at War.* Ruskins Publishing, Halesworth
Bolitho H. 1929. *The Glorious Oyster.* Sidgwick and Jackson, London
Bragg B. 2006. A13: Trunk road to the sea. V & A Memory maps. At www.vam.ac.uk/activ_events/adult_resources/memory_maps
Bruce K. 1981. *Dengie. The Life and the Land.* Essex Records Office, Chelmsford
Clegg R. and Clegg S. 1999. *Southwold.* Phillimore, Chichester
Cobb D. 2006. *Business in Eden.* Equinox Press, Braintree
Cracknell B E. 1910. *Canvey Island. Captivating Canvey Guides.* University Press, Leicester
Crossley A., Dunn M. and Heard F. 2000. *Tollesbury in the Year 2000.* Tollesbury Millennium Publ, Tollesbury
Currie I, Davison M. and Ogilvy B. 1992. *The Essex Weather Book.* Froglets Publ, Westerham
Defoe D. 1724–26. [1972]. *A Tour Through the Whole Island of Great Britain.* Penguin, London
Dutt W A. 1904. *Highways and Byways in East Anglia.* MacMillan, London
Dutt W. 1910. *The Norfolk and Suffolk Coast.* Frederick Stokes, New York
Evans G. E. 1955. *Ask the Fellows Who Cut the Hay.* Faber & Faber, London
Evans G. E. 1960. *The Horse and the Furrow.* Faber & Faber, London
Evans G. E. 1975. *The Days That We Have Seen.* Faber & Faber, London
Evans G. E. 1979. *Horse Power and Magic.* Faber & Faber, London

Gould I. and Gould B. M. 1911. *An Anthology of Essex.* Samson Low Marston and Co, London

Grieve H. 1959. *The Great Tide.* Essex County Council, Chelmsford

Harman R. (ed). 1949. *Country Company.* Blandford Press, London

Harrison L. 2007. *Captivating Canvey.* Rendezvous Press, London

James H. 1905 [1981]. *English Hours.* Oxford Paperback, Oxford

Jarvis S. 1993. *Essex Headlines.* Countryside Books, Newbury

Jarvis S. 1993. *Essex. A County History.* Countryside Books, Newbury

Johnson D. E. 1992. *East Anglia at War 1939-1945.* Jarold, Norwich

Macfarlane R. 2008. Ghost species. In Granta 102. *The New Nature Writing.* Granta, London

Maxwell D. 1925. *Unknown Essex.* Bodley Head, London

Mee A. 1941. *Suffolk: The King's England.* Hodder and Stoughton, London

Mooney B. and Harris J. 2004. *Frontier Country: A Mid-Summer Walk Around the Borders of Essex.* Thorogood, London

Morrison A. 1900. *Cunning Murrell.* Doubleday, Page and Co, New York

Motley N. 1985. *Much Ado About Something.* Othona Community, Bradwell

Newell M. 1999. *Black Shuck.* Jardine Press, Hadleigh

Orton J. and Worpole K. 2005. *350 Miles. An Essex Journey.* Essex Development and Regeneration Agency, Chelmsford

Pewsey L. 1997. *A Taste of Essex.* Ian Henry Press, Romford

Ransome A. 1969. *Secret Water.* Puffin Books, London

Roberts M. and Roberts R. 2006. *Paglesham Natives.* M. A. and R. Roberts, St Osyth

Rhys Jones G. 2006. *Semi-Detached.* Penguin, London

Rider Haggard H. 1901 and 1902. *Rural England.* Vols 1 and 2. Adamant Media [2000, 2001], Boston

Runcie J. 2006. *Canvey Island.* Bloomsbury, London

Salmon J. 1977. *The Suffolk and Essex Border.* Boydell Press, Ipswich

Scarfe N. 1960. *Essex.* Faber & Faber, London

Scarfe N. 1960. *Suffolk.* Faber & Faber, London

Sebald W. G. 2002. *The Rings of Saturn.* Vintage, London

Smith G. 2005. *Smuggling in Essex.* Countryside Books, Newbury

Smith J. R. 1970. *Foulness. A History of an Essex Island Parish.* Essex County Council, Chelmsford

Sorrell M. 1977. *The Peculiar People.* Paternoster Press, Exeter

Sparrow W. 2006. *Memories of Nayland in the 20th Century.* Nayland and Wissington Conservation Society

Storey N. R. 2006. *The Lost Coast of Norfolk.* Sutton, Stroud

Strachan D. 1998. *Essex from the Air.* Essex County Council, Chelmsford

Suffolk Federation of WIs. 1994. *Suffolk: Within Living Memory.* Countryside Books, Newbury

Swanton M. 2000. *The Anglo-Saxon Chronicles.* Phoenix Press, London

Taylor A. R. and Tonks E. S. 1979. *Southwold Railway.* Ian Allan, London

Tompkins H. 1904. *Marsh Country Rambles.* Chatto and Windus, London

Walker D. 2001. *Basildon Plotlands.* Phillimore, Chichester

Waller A. J. R. 1959. *The Suffolk Stour.* Norman Alard, Ipswich

Ward Lock. 1950. *Guide to Felixstowe and the Suffolk Coast.* London

Wentworth Day J. 1941. *Farming Adventure.* Harrap, London

Wentworth Day J. 1979. *Book of Essex.* Egon, Letchworth

Williamson T. 2006. *England's Landscapes. East Anglia.* Collins, London

Worpole K. 2006. *Estuary Lines.* At V & A Memory Maps. URL www.vam.ac.uk/activ_events/adult_resources/memory_maps

Young A. 1813. *General View of the Agriculture of the County of Suffolk.* David and Charles [1969], Plymouth

Acknowledgements

I am grateful to the many people who helped in the organisation of these coastal walks and journeys, to those who accompanied me, and to those who generously spared the time to talk and navigate me through their places, memories and lives. Others gave advice on walking, story-telling, writing, natural and social history, and on passing through landscapes. Some held doors open, and others gave lifts. I hope I told their stories with fairness and accuracy. Very many thanks to:

Will Akast, Peter Avery, Chris Barningham, Jo Barton, Ted Benton, Gill Boardman, Ronald Blythe, Peter Boggis, Mark Bridges, Elizabeth Buchanan, Philippa Buckley, Anne and John Chittock, Matt Cole, Steve Cooper and Jeanette Johnson, Edward Cross, Brian Dawson, Matthew Dell, Roy and Linda Ducker, James Dudderidge, Gerald Engwell, Peter Ennis, Fred Farenden, Lee Firmin, Sally Green, Jilly Hall, John Hall, Richard Haward, Jeff Harrison, Lucy Harrison, Gail Hayne, Lorna Heffron, Bob Hills, Rachel Hine, Brian Hodgson, Simon Hooton, Amanda Hopkinson, Rene Horwood, Matt Hudson, Mary Hull, Peter Hulme, Brian Johnson, Liam Kelly, William Kendall, Malcolm Kerby, Simon Loftus, Grant Lohoar, Derek and Valerie Lynch, Ryan Lynch, Richard Mabey, Rob Macfarlane, John Marsh, Ray Marsh, Andy May, Bill McKibben, Peter Melchett, Leo Mellor, Dom Micklewright, Jeremy Millar, Rick Minter, Simon Moore, Marcus O'Lone, Justin Partyka, Andrew Phillips, Sarah Pilgrim, Chris Pretty, John and Susan Pretty, Marilyn Pritchard, Michael Pudney, Amanda Raithby, David Reed, Adam Rochester, Julian Roughton, Steven Saint, Mike Sandison, Sam and Mary Self, Martin Sellens, Alan Shearring, Dave Smart, Richard Starling, Phil Sturges, Ray Tabor, Janet and Pip Thorogood, Paul Trathen, Nicki Uden, Graham Underwood, Bryan Upson, Deanna Walker, Janie Walker, Marina Warner, David Williams, Leon Woodrow, Ken Worpole, Eric Wortley, Stephen and Peter Wortley.

I am particularly thankful to Glenn Albrecht, Louis Baum, Gill Boardman, Madeleine Bunting, Liz Calder, John and Genevieve Christie, Jenny Harpur, Peter Hulme, Richard Mabey, Rob Macfarlane, Tero Mustonen, Gwen Roland, and Ken Worpole for their acute comments and sage advice during editing and production. Any errors and misjudgements are mine alone.

A NOTE ON WEIGHTS AND MEASURES

As these were contemporary journeys as well as descents into history, I have allowed weights and measures to be both imperial and metric. Mostly distances are in miles, but sometimes kilometres, and on a smaller scale, variously feet, yards and metres; areas in hectares, but acres where local people still think that way; weight in kilograms and pounds, tonnes and also tons; volumes in litres and gallons; and speed in miles or kilometres per hour, or knots at sea. These renamings of terms, and the subsequent and inevitable confusion, echo many of the changes in name and identity that places and people on the coast have undergone.

First published in 2011 by Full Circle Editions

Parham House Barn, Brick Lane, Framlingham, Woodbridge, Suffolk IP13 9LQ
www.fullcircle–editions.co.uk

 A CIP record for this book is available from the British Library.

Set in Bell MT & Gill Sans
Paper: Munken Pure 120gsm from FSC Mixed Sources

Book design: Jonathan Christie
Printed and bound in Suffolk by Healeys Print Group, Ipswich.

ISBN 978-0-9561869-6-6

Note on the typeface:
Bell MT was made in 1931 by Monotype as a facsimile of the typeface cut originally for John Bell by Richard Austin in 1788. Taking the matrices in the possession of Stephenson Blake & Co as a basis, it was used in John Bell's newspaper, "The Oracle," and was regarded by Stanley Morison as the first English Modern face. Although inspired by French punchcutters of the time, having a vertical stress and fine hairlines, the face is less severe than the French models and is now classified as Transitional.

LOTTERY FUNDED

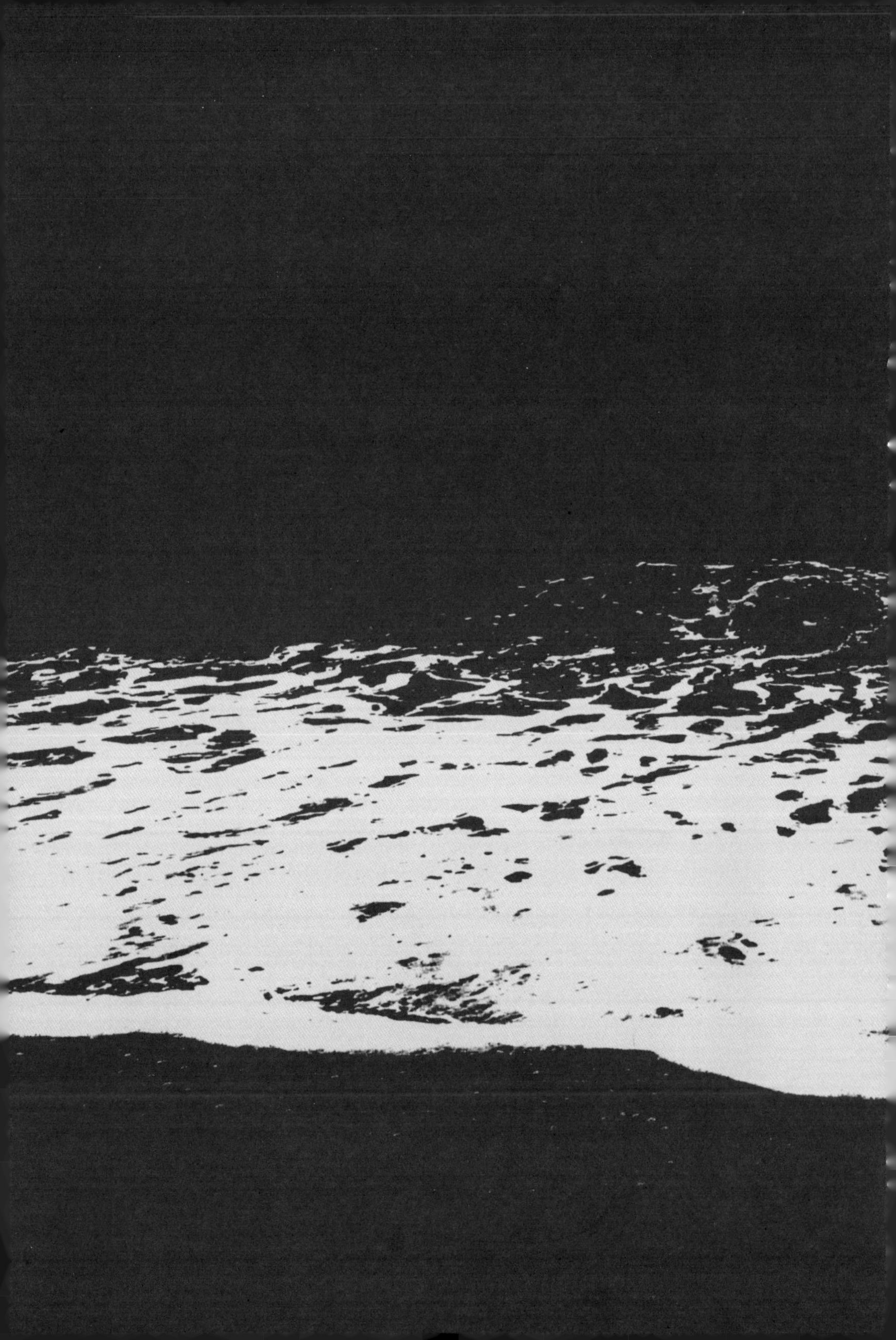